T0249038

IET SECURITY SERIES 2

Engineering Secure Internet of Things Systems

Engineering Secure Internet of Things Systems

Edited by Benjamin Aziz,
Alvaro Arenas and Bruno Crispo

The Institution of Engineering and Technology

Published by The Institution of Engineering and Technology, London, United Kingdom

The Institution of Engineering and Technology is registered as a Charity in England & Wales (no. 211014) and Scotland (no. SC038698).

First published 2016

The Institution of Engineering and Technology
Michael Faraday House
Six Hills Way, Stevenage
Herts, SG1 2AY, United Kingdom

www.theiet.org

British Library Cataloguing in Publication Data
A catalogue record for this product is available from the British Library

ISBN 978-1-78561-053-0 (hardback)
ISBN 978-1-78561-054-7 (PDF)

Typeset in India by MPS Limited

Contents

Preface

Benjamin Aziz, Alvaro Arenas and Bruno Crispo

Overview

The Internet of Things (IoT) is a new emerging paradigm that refers to the connectivity of "anything" that can carry a minimum amount of data storage and computational power (e.g. cyber-physical systems) to the Internet, therefore promoting collaboration and knowledge anytime anywhere in order to improve the quality of everyday life and open up new business opportunities. Recent years have witnessed the emergence of global standardisation bodies such as the International Telecommunication Union's "Internet of Things Global Standards Initiative" as well as communication standards such as one Machine-to-Machine (M2M), MQ Telemetry Transport (MQTT), Common Opensource Application Publishing Platform (CoAPP) and eXtensible Messaging and Presence Protocol (XMPP). These efforts have led to an increasing role and usage of IoT systems in modern societies, the economies they support and infrastructures they are based on. Nowadays, IoT applications occupy almost every domain of life; for example, personal health care, transportation and vehicular connectivity, industrial embedded systems, enterprise businesses using wireless sensor networks and any services with utility monitoring as well as traditional areas of avionics, space industries and national defence.

This increasing *globalisation* of IoT applications has led to rising concerns about questions related to how private data collected by such applications will be protected and what levels of assurance can one maintain with regards to the reliability, security and trustworthiness of these applications. One answer to these concerns is to integrate the body of wisdom that the scientific research community has developed over the decades, in areas of computing reliability, security and trust, into the software engineering process of IoT systems. The current book represents a small step towards this integration effort that we hope will impact the security and robustness of future IoT systems, applications and standards in a positive manner.

The book presents a collection of state-of-the-art research essays in themes related to areas of privacy, identity and authorisation management, formal security modelling and analysis, cryptography-based solutions and trust and reputation management in the context of IoT systems. As a result, the book covers a wide range of issues currently at the forefront of the security concerns in the IoT. The book will benefit primarily academic researchers and researchers working in industry who are interested in the topic of security in the context of IoT systems. The book will also be of interest to professional managers and members of technical standardisation bodies with scientific background.

Book structure

The book consists of ten chapters. We next provide an overview of these.

A security survey of middleware for the Internet of Things

This chapter presents a thorough survey of the current state-of-the-art middleware systems for the IoT and discusses the various security and privacy challenges that such systems both raise and help to solve. IoT middleware systems constitute the *fabric* of the IoT over which business applications and solutions are developed. Therefore, understanding the current security and privacy challenges in this fabric is crucial to the understanding of the reliability and robustness of any solutions that are implemented over it. The chapter concludes with the identification of gaps in security and privacy-related research in IoT middleware systems and makes recommendations for future work that can fulfil these gaps.

Privacy in the Internet of Things

Of particular importance is the privacy of users when it comes to the usage of IoT devices, given their ever-increasing incorporation in modern every-day-life devices. One of the long-standing arguments in the privacy community is that privacy should be *by design* rather than an afterthought. As a result, this chapter focuses on treating privacy in IoT systems by proposing a "Privacy Development Life Cycle" that incorporates the privacy concern throughout the whole software/system development life cycle. The chapter proposes a number of PETs (Privacy Enhancing Technologies) for the IoT, and ends with a conclusion that existing PETs developed for non-IoT systems require more efficiency tuning to render them suitable for the constrained nature of computational environments within which IoT systems often run. Overall, the chapter provides a view on why the IoT both increases the likelihood and aggravates the impact of privacy breaches in Information Technology (IT) systems.

Privacy and consumer IoT: a sensemaking perspective

The success of any system is ultimately down to its ability to address consumer needs and worries, and its ability to create value where new business opportunities emerge. The IoT is not different in this respect, and this chapter aims at addressing the role of various stakeholders in an IoT ecosystem using *sensemaking theory* paying particular attention to the privacy concern of consumers within such an ecosystem. The chapter covers literature at the intersection of the sensemaking theory and the IoT, and further proposes an adaptation of the sensemaking theory to fill the gap between users' and provider's aspirations of control of information generated within IoT systems.

SMARTIE: a secure platform for Smart Cities and IoT

Whilst security and privacy in the context of IoT systems is treated from a high-level development life cycle and a holistic sensemaking perspective in the previous two chapters, this chapter presents a more concrete architectural view of how these issues can be treated in an IoT system. The chapter provides an overview of a high-level architecture reference model for IoT systems, which was developed within a recent European project called SMARTIE. The reference model incorporates functional components that care of security and privacy concerns. The applicability of this architecture is demonstrated through various use cases, such as smart energy management, public transport networks, urban traffic management infrastructures and smart city information centres. The reference model is defined based on *functional groups* of components, which means it can be implemented with any technology that provides the required interface of functionality. These groups include elements such as distributed capability-based access control, XACML, Elliptic Key Cryptography, Kerberos and others.

Model-based security engineering for the Internet of Things

The concept of a model-driven architecture is based on the idea that the architecture's design and specification can be generated automatically directly from a specification of the requirements that underlie the business application(s), for which the system is being developed, as well as the constraints and limitations posed by the technologies used for its development. Requirements are usually specified in some input domain-specific language and are then transformed into enforceable rules, mechanisms, configurations and analyses applied to the system's design and specification with minimum human intervention. This chapter addresses the need for a *model-driven security engineering* approach in the context of existing complex IoT systems, by proposing the model-based Security toolKit (SecKit) and framework that can be used to express and enforce security requirements. The framework addresses requirements related to the structure, behaviour, identity definition, business roles and contextual information for an IoT system. The applicability of framework is demonstrated through a business case study related to the real-time monitoring of transportation conditions of perishable goods, also known as the *cold chain monitoring* problem.

Federated Identity and Access Management in IoT systems

One of the most challenging issues in building any large-scale computing system is the issue of managing identities across multiple domains and platforms, and enforcing good access control decisions for based on such identities. This problem is also referred to as *Federated Identity and Access Management*, and it constitutes the focus of the next chapter, which considers how the federated identity and access management problem can be solved in an IoT system by translating existing Web-based standards for the management of federated identities, in particular OAuth2, to the world of IoT systems. The chapter outlines one implementation of this approach for the case of the MQTT protocol and based on computationally constrained devices, in this case Arduino boards. The chapter also evaluates the performance of the resulting system and discusses the pros and cons of this approach in solving the IoT Federated Identity and Access Management problem.

On the security of the MQTT protocol

Developing complex secure systems involves a fundamental design-time stage; that of the analysis of the formal specifications and standards underlying the system's design and architecture. Such formal analysis, also known as Verification and Validation (V&V), is important particularly where the system under development contains critical components whose failure would be highly risky for the rest of the system. Despite the cost and time often involved in the V&V stage of the software development life cycle, it remains an essential exercise due to its role in uncovering high-risk deficiencies and vulnerabilities, which can undermine a system's reliability and security properties, to a high level of accuracy and precision. Following this approach, the chapter provides a formal analysis of an earlier version of the MQTT protocol (v. 3.1), which uncovers an example of semantic ambiguities only possible to detect using a formal analysis approach but that could lead to weak implementations of MQTT-based systems. The chapter also discusses how the uncovered ambiguity can lead to security threats by various types of attackers in an IoT network.

Securing communications among severely constrained, wireless embedded devices

The proliferation of low-cost low-computational power devices in recent years has created a need for the development of lightweight communication protocols (e.g. MQTT, CoAPP and others) that can run over unreliable communication networks in an efficient and simple manner within and across an IoT ecosystem. This need is demonstrated even more in commercial scenarios, such as smart cities' infrastructures, where the applications often run there involve rich requirements both on the business and security/privacy levels. Cryptography remains an essential element of ensuring that such security/privacy-relevant requirements are met. Therefore, this chapter presents a new framework, developed within the European project RERUM, for constructing lightweight cryptographic solutions for IoT communications running at the datagram transport layer. The framework is based on the theory of *compressive sensing* and takes into account the integrity verification of data using on-device signatures, as an example of security-relevant requirements in a smart city scenario.

Lightweight cryptographic identity solutions for the Internet of Things

The same problem discussed in the previous chapter, i.e. that of ensuring that secure communications among IoT devices with low-computational power are implemented in an efficient manner, is tackled again in this chapter, however this time from a hardware point of view. The chapter presents a solution based on the concept of *physical unclonable functions*. Physical unclonable functions are the physical representation of one-way functions that are easy to manufacture within an IoT device and exhibit two important properties: first, they are easy to compute but hard to predict, and second, they are hard to duplicate even if their exact manufacturing conditions are known. The chapter demonstrates an example of how such functions can be manufactured and applied to securing communications among devices with low-computational power.

A reputation model for the Internet of Things

The final chapter in this book considers another aspect of securing IoT systems, that of trust management based on reputation of how well the devices, servers and applications in an IoT system behave over time. The chapter first presents a reputation model for IoT networks based on the concept of *utility functions*, which provide a measure of the satisfaction of an entity with regards to the quality of service it receives from another entity. Reputation is measured with respect to some *issue of interest*, and in this case, the issue of interest considered is that of the number of times a message is delivered to an application by the IoT device and server. The chapter concludes with the presentation of an architecture for implementing a reputation system in an IoT network and evaluates the results obtained based on one implementation of this architecture, which applied the reputation model, against the expected theoretical results.

Chapter 1

A security survey of middleware for the Internet of Things

Paul Fremantle

Summary

The rapid growth of small Internet connected devices, known as the Internet of Things (IoT), is creating a new set of challenges to create secure, private infrastructures. The purpose of this chapter is to review the current literature on the challenges and approaches to security and privacy in the IoT, with an especial focus on how these aspects are handled in IoT middleware. We focus on IoT middleware because many systems are built from existing middleware and these inherit the underlying security properties of the middleware framework.

The chapter is composed of three main sections. First, we look at the general security and privacy challenges around IoT. Second, we present a structured literature review of the available middleware and how security is handled in these middleware approaches. Finally, we draw a set of conclusions and identify further work in this area.

1.1 Introduction

The *Internet of Things* (*IoT*) was originally coined as a phrase by Kevin Ashton in 1990 [1], with reference to "taggable" items that used Radio-frequency identification (RFID) chips to become electronically identifiable and therefore amenable to inter-actions with the Internet. With the ubiquity of cheap processors and System-on-Chip (SoC)-based devices, the definition has expanded to include wireless and Internet-attached sensors and actuators, including smart meters, home automation systems, Internet-attached set-top boxes, smartphones, connected cars, and other systems that connect the physical world to the Internet either by measuring it or affecting it.

The IoT is defined variously, but we will use the definition from Reference 2:

"A dynamic global network infrastructure with self configuring capabilities based on standard and interoperable communication protocols where physical and virtual 'things' have identities, physical attributes, and virtual personalities and use intelligent interfaces, and are seamlessly integrated into the information network."

The number of IoT devices has grown rapidly, with a recent estimate suggesting that there were 12.5 billion Internet-attached devices in 2010 and a prediction of 50 billion devices by 2020 [3]. This brings with it multiple security challenges:

- These devices are becoming more central to people's lives, and hence the security is becoming more important.
- These devices, due to size and power limitations, may not support the same level of security that we would expect from more traditional Internet-connected devices.
- The sheer scale and number of predicted devices will create new challenges and require new approaches to security.

We will explore each of these areas in more detail below, as we look at the potential attacks that are possible on IoT devices.

1.2   Approach

In order to understand the security threats against the IoT, we need to take an approach to classifying threats. The most widely used ontology of security threats is based on the classic "CIA" (Confidentiality, Integrity, Availability) model [4] which has been extended over the years and is now often referred to as the "CIA+" model [5]. In the course of reviewing the available literature and approaches to IoT security, we have created a proposed expansion of the existing framework that we believe works better in the IoT space. In particular, we propose a new ontology based on a matrix of evaluation where we look at each of the classic security challenges in three different aspects: device/hardware, network, and cloud/server-side. In some cells in this matrix, we have not identified any areas where the IoT space presents new challenges; in other words, while the domain space covered by these cells contains security challenges, those challenges are no different from existing Web and Internet security challenges in that domain. In those cells, we can say that the challenges are "unchanged". In other cells, we specifically identify those challenges that are significantly modified by the unique nature of the IoT.

Figure 1.1 shows the matrix we will use for evaluating security challenges. In each cell, we summarise the main challenges that are *different* in the IoT world or at least exacerbated by the challenges of IoT compared to existing Internet security challenges. We will explore each cell in the matrix in detail below. Each of the cells is given a letter from *A* to *R* and these letters are used as a key to refer to the cells below.

The three aspects (hardware/device, network, cloud/server) were chosen because as we read the available literature these areas became clear as a way of segmenting the unique challenges within the context of the IoT compared with existing Internet security challenges. These form a clear logical grouping of the different assets involved in IoT systems. We will provide a quick overview of each area before we look in detail at each cell of the matrix.

Device and hardware
IoT devices have specific challenges that go beyond those of existing Internet clients. These challenges come from the different form factors of IoT devices,

Security characteristic	Device/hardware	Network	Cloud/server-side
Confidentiality	A. Hardware attacks	B. Encryption with low capability devices	C. Privacy concerns
Integrity	D. Lack of attestation, illicit updates	E. Signatures with low capability devices	F. Unchanged
Availability	G. Physical attacks; radio jamming	H. Unreliable networks	I. Unchanged
Authentication	J. Lack of user input; hardware retrieval of keys	K. Challenges of using federated identity	L. Lack of widely implemented standards around device identity
Access control	M. Physical access; lack of local authentication	N. Lightweight protocols for access control	O. Requirement for user-managed access controls
Non-repudiation	P. No secure local storage; low capability devices	Q. Signatures with low capability devices	R. Unchanged

Figure 1.1   Matrix of security challenges for the IoT

from the power requirements of IoT devices, and from the hardware aspects of IoT devices. The rise of cheap mobile telephony has driven down the costs of 32-bit processors (especially those based around the Advanced RISC Machine (ARM) architecture [6]), and this is increasingly creating lower cost microcontrollers and SoC devices based on ARM. However, there are still many IoT devices built on 8-bit processors, and occasionally, 16-bit [7]. In particular, the open-source hardware platform Arduino [8] supports both 8-bit and 32-bit controllers, but the 8-bit controllers remain considerably cheaper and more popular.

Network

IoT devices may use much lower power, lower bandwidth networks than existing Internet systems. Cellular networks often have much higher latency and more "dropouts" than fixed networks [9]. The protocols that are used for the Web are often too data-intensive and power-hungry for IoT devices. Network security approaches such as encryption and digital signatures are difficult or impractical in small devices.

Cloud/server-side

While many of the existing challenges apply here, there are some aspects that are exacerbated by the IoT for the server-side or cloud infrastructure. These include: the often highly personal nature of data that is being collected and the requirement to manage privacy; the need to provide user-managed controls for access; and the lack of clear identities for devices making it easier to spoof or impersonate devices.

1.2.1   *Cell A: confidentiality/hardware*

The confidentiality of data on the device itself can certainly be an issue. While it could be argued that many IoT devices are in public areas, even these devices may store historical data locally, or may have security data (e.g. keys, passwords, credentials or sensitive code) that is liable to attack. Reference 10 is a comprehensive study of many semi-invasive attacks that can be done on hardware. References 11, 12 cover the concept of *side channel attacks* where the power usage or other indirect information from the device can be used to steal information.

A related issue to confidentiality of the data on the device is the challenges inherent in updating devices and pushing keys out to devices.

The distribution and maintenance of certificates and public keys onto embedded devices are complex [13]. In addition, sensor networks may be connected intermittently to the network resulting in limited or no access to the Certificate Authority (CA). To address this, the use of threshold cryptographic systems that do not depend on a single central CA has been proposed [14], but this technology is not widely adopted: in any given environment this would require many heterogeneous Things to support the same threshold cryptographic approach.

We can also see clearly from a number of publicised attacks [15–17] that device designers have not adjusted to the challenges of designing devices that will be connected either directly or indirectly to the Internet.

A further security challenge for confidentiality and hardware is the fingerprinting of sensors or data from sensors. In Reference 18, it has been shown that microphones, accelerometers, and other sensors within devices have unique "fingerprints" that can uniquely identify devices.

Finally, the use of Public Key Infrastructure (PKI) requires devices to be updated as certificates expire. The complexity of performing updates on IoT devices is harder, especially in smaller devices where there is no user interface. For example, many devices need to be connected to a laptop in order to perform updates. This requires human intervention and validation, and in many cases this is another area where security falls down. For example, many situations exist where security flaws have been fixed but because devices are in homes, or remote locations, or seen as appliances rather than computing devices, updates are not installed [17].

1.2.2   *Cell B: confidentiality/network*

The confidentiality of data on the network is usually protected by encryption of the data. There are a number of challenges with using encryption in small devices. Performing public-key encryption on 8-bit microcontrollers has been enhanced by the use of Elliptical Curve Cryptography (ECC) [19, 20]. ECC reduces the time and power requirements for the same level of encryption as an equivalent Rivest Shamir Adleman (RSA) public-key encryption [21] by an order of magnitude [22–24]; RSA encryption on constrained 8-bit microcontrollers may take minutes to complete, whereas similar ECC-based cryptography completes in seconds. However, despite the fact that ECC enables 8-bit microcontrollers to participate in public-key encryption systems, in many cases it is not used. We can speculate as to why this is: first, as evidenced by Sethi *et al.* [24], the encryption algorithms consume a large proportion of the available Read-Only-Memory (ROM) on small controllers. Second, there is a lack of standard open-source software. For example, a search that we carried out (on the 21 April 2015) of the popular open-source site Github for the words "Arduino" and "Encryption" revealed 10 repositories compared to "Arduino" and "HTTP" which revealed 467 repositories. These 10 repositories were not limited to network level encryption. However, recently an open-source library for Advanced Encryption Standard (AES) on Arduino [25] has made it more effective to use cryptography on Atmel-based hardware.

Another key challenge in confidentiality is the complexity of the most commonly used encryption protocols. The standard Transport Layer Security (TLS [26]) protocol can be configured to use ECC, but even in this case the handshake process requires a number of message flows and is sub-optimal for small devices as documented in Reference 27. Perelman and Ersue [28] has argued that using TLS with *Pre-Shared Keys* (PSK) improves the handshake. However, they fail to discuss in any detail the significant challenges with using PSK with IoT devices: the fact that either individual symmetric key need to be deployed onto each device during the device manufacturing process, or the same key re-used. In this case, there is a serious security risk that a single device will be broken and thus the key will be available.

There is an alternative protocol for UDP: DTLS (Datagram Transport Level Security) [29] which provides a lighter weight approach than TLS. However, there is still a reasonably large Random Access Memory (RAM) and ROM size required for this [30], and this requires that messages be sent over UDP which has significant issues with firewalls and home routers, making it a less effective protocol for IoT applications [31]. There is ongoing work at the Internet Engineering Task Force (IETF) to produce an effective profile of both TLS and DTLS for the IoT [32].

A significant area of challenge for network confidentiality in IoT is the emergence of new radio protocols for networking. Previously, there were equivalent challenges with Wi-Fi networks as protocols such as WEP were broken [33], and there are new attacks on protocols such as Bluetooth 4.0 (also known as Bluetooth LE/BLE). For example, while BLE utilises AES encryption which has a known security profile, a new key exchange protocol was created, which turns out to be flawed, allowing any attacker present during key exchange to intercept all future communications [34]. One significant challenge for IoT is the length of time it takes for vulnerabilities to be addressed when hardware assets are involved. While the BLE key exchange issues are addressed in the latest revision of BLE, we can expect it to take a very long time for the devices that encode the flawed version in hardware to be replaced, due to the very large number of devices and the lack of updates for many devices. By analogy, many years after the WEP issues were uncovered, in 2011 a study showed that 25% of Wi-Fi networks were still at risk [35].

In Reference 36, a theoretical model of traceability of IoT devices and particularly RFID systems are proposed in order to prevent unauthorised data being accessible. A protocol that preserves the concept of untraceability is proposed.

Many of the same references and issues apply to section *E* where we look at the use of digital signatures with low power devices.

1.2.3   Cell C: confidentiality and cloud/server

While the issues here are largely similar to normal Web- and Internet-based systems, there are certainly concerns about individual's privacy on with the IoT. For example, the company Fitbit [37] made data about users sexual activity available and easily searchable online [38] by default. There are social and policy issues regarding the ownership of data created by IoT devices [39, 40]. We address these issues in more detail in the Cell *O* where we look at the access control of IoT data and systems in the cloud and on the server-side.

A second concern that is exacerbated by the IoTs are concerns with correlation of data and metadata, especially around *de-anonymisation*. In Reference 41, it was shown that anonymous metadata could be de-anonymised by correlating it with other publicly available social metadata. This is a significant concern with IoT data. This is also closely related to the fingerprinting of sensors within devices as discussed in Cell *A*. An important model for addressing these issues in the cloud are systems that filter, summarise, and use *stream-processing* technologies to the data coming from IoT devices before this data is more widely published. For example, if we only publish a summarised co-ordinate rather than the raw accelerometer data, we can potentially avoid fingerprinting de-anonymisation attacks.

In addition, an important concern has been raised in the recent past with the details of the government sponsored attacks from the US National Security Agency (NSA) and British Government Communications Headquarters (GCHQ) that have been revealed by Edward Snowden [42]. These bring up three specific concerns on IoT privacy and confidentiality.

The first concern is the revelations that many of the encryption and security systems have had deliberate backdoor attacks added to them so as to make them less secure [43]. The second concern is the revelation that many providers of cloud hosting systems have been forced to hand over encryption keys to the security services [44]. The third major concern is the revelations on the extent to which metadata is utilised by the security services to build up a detailed picture of individual users [45].

The implications of these three concerns when considered in the light of the IoT are clear: a significantly deeper and larger amount of data and metadata will be available to security services and to other attackers who can utilise the same weaknesses that the security services compromise.

1.2.4   Cell D: integrity and hardware/device

The concept of integrity refers to maintaining the accuracy and consistency of data. In this cell of the matrix, the challenges are in maintaining the device's code and stored data so that it can be trusted over the life cycle of that device. In particular, the integrity of the code is vital if we are to trust the data that comes from the device or the data that is sent to the device. The challenges here are viruses, firmware attacks, and specific manipulation of hardware. For example, Goodin [46] describes a worm attack on router and IoT firmware.

The traditional solution to such problems is attestation [47–49]. Attestation is important in two ways. First, attestation can be used by a remote system to ensure that the firmware is unmodified and therefore the data coming from the device is accurate. Second, attestation is used in conjunction with hardware-based secure storage (Hardware Security Managers, as described in Reference 50) to ensure that authentication keys are not misused. The model is as follows.

In order to preserve the security of authentication keys in a machine where human interaction is involved, the user is required to authenticate. Often the keys are themselves encrypted using the human's password or a derivative of the identification parameters. However, in an unattended system, there is no human interaction. Therefore, the authentication keys need to be protected in some other way. Encryption on its own is no help, because the encryption key is then needed and this becomes a circular problem. The solution to this is to store the authentication key in a dedicated hardware storage. However, if the firmware of the device is modified, then the modified firmware can read the authentication key, and offer it to a hacker or misuse it directly. The solution to this is for an attestation process to validate the firmware is unmodified before allowing the keys to be used. Then the keys must also be encrypted before sending them over any network.

These attestation models are promoted by groups such the Trusted Computing Group [51] and Samsung Knox [52]. These rely on specialised hardware chips such as the Atmel AT97SC3204 which implement the concept of a Trusted Platform Module (TPM) [53]. There is research into running these for Smart Grid devices [54]. However, while there is considerable discussion of using these techniques with IoT, during our literature review, we could not find evidence of any real-world devices apart from those based on mobile phone platforms (e.g. phones and tablets) that implemented trusted computing and attestation.

1.2.5 Cell E: integrity and network

Maintaining integrity over a network is managed as part of the public-key encryption models by the use of digital signatures. The challenges for IoT are exactly those we already identified in the section B above where we described the challenges of using encryption from low-power IoT devices.

1.2.6 Cell F: integrity and cloud/server

This area is unchanged by the IoT, so we do not discuss it any further.

1.2.7 Cell G: availability and device/hardware

One of the significant models used by attackers is to challenge the availability of a system, usually through a Denial of Service (DoS) or Distributed Denial of Service (DDoS) attack. DoS attacks and availability attacks are used in several ways by attackers. First, there may be some pure malicious or destructive urge (e.g. revenge, commercial harm, share price manipulation) in bringing down a system. Second, availability attacks are often used as a pre-cursor to an authentication or spoofing attack.

IoT devices have some different attack vectors for availability attacks. These include resource consumption attacks (overloading restricted devices), physical attacks on devices.

1.2.8   Cell H: availability and network

There are clearly many aspects of this that are the same as existing network challenges. However, there are some issues that particularly affect IoT. In particular, there are a number of attacks on local radio networks that are possible. Many IoT devices use radio networking (Bluetooth, Wi-Fi, 3G, General Packet Radio Service (GPRS), LoRa, and others) and these can be susceptible to radio jamming. Another clear area of attack is simply physical access. For example, even wired networks are much more susceptible to DoS attacks when the devices are spread widely over large areas.

1.2.9   Cell I: availability and cloud/server

This area is unchanged by the IoT, so we do not discuss it any further.

1.2.10   Cell J: authentication and device/hardware

We will consider the authentication of the device to the rest of the world in later sections. In this cell of the matrix, we must consider the challenges of how users or other devices can securely authenticate to the device itself. These are however related: a user may bypass or fake the authentication to the device and thereby cause the device to incorrectly identify itself over the network to other parts of the Internet.

Some attacks are very simple: many devices come with default passwords which are never changed by owners. In a well-publicised example [17], a security researcher gained access to full controls of a number of "smart homes".

Similarly many home routers are at risk through insecure authentication [55]. Such vulnerabilities can then spread to other devices on the same network as attackers take control of the local area network.

A key issue here is the initial *registration* of the device. A major issue with hardware is when the same credential, key, or password is stored on many devices. Devices are susceptible to hardware attacks (as discussed above) and the result is that the loss of a single device may compromise many or all devices. In order to prevent this, devices either must be pre-programmed with unique identifiers and credentials at manufacturing time, or must go through a registration process at set-up time. In both cases, this adds complexity and expense, and may compromise usability. We argued for the use of the Dynamic Client Registration process in Reference 56 to create unique keys/credentials for each device.

1.2.11   Cell K: authentication and network

Unlike browsers or laptops where a human has the opportunity to provide authentication information such as a user ID and password, IoT devices normally run unattended and need to be able to power cycle and reboot without human interaction. This means that any identifier for the device needs to be stored in the program memory (usually Static RAM (SRAM)), ROM or storage of the device. This brings two distinct challenges:

- The device may validly authenticate, but its code may have been changed.
- Another device may steal the authentication identifier and may *spoof* the device.

In the Sybil attack [57], a single node or nodes may impersonate a large number of different nodes thereby taking over a whole network of sensors.

In all cases, attestation is a key defence against these attacks.

Another defence is the use of *reputation* and reputational models to associate a trust value to devices on the network.

Reputation is a general concept widely used in all aspects of knowledge ranging from humanities, arts, and social sciences to digital sciences. In computing systems, reputation is considered as a *measure* of how trustworthy a system is. There are two approaches to trust in computer networks: the first involves a "black and white" approach based on security certificates, policies, etc. For example, SPINS [58] develops a trusted network. The second approach is probabilistic in nature, where trust is based on reputation, which is defined as a probability that an agent is trustworthy. In fact, reputation is often seen as one measure by which trust or distrust can be built based on good or bad past experiences and observations (direct trust) [59] or based on collected referral information (indirect trust) [60].

In recent years, the concept of reputation has shown itself to be useful in many areas of research in computer science, particularly in the context of distributed and collaborative systems, where interesting issues of trust and security manifest themselves. Therefore, one encounters several definitions, models and systems of reputation in distributed computing research (e.g. References 59, 61, 62).

There is considerable work into reputation and trust for wireless sensor networks, much of which is directly relevant to IoT trust and reputation. The Hermes and E-Hermes [63, 64] systems utilise Bayesian statistical methods to calculate reputation based on how effectively nodes in a mesh network propagate messages including the reputation messages. Similarly, Chen *et al.* [65] evaluate reputation based on the packet-forwarding trustworthiness of nodes, in this case using fuzzy logic to provide the evaluation framework. Another similar work is [66] which again looks at the packet forwarding reputation of nodes.

1.2.12 Cell L: authentication and cloud/server

The IETF has published a draft guidance on security considerations for IoT [67]. This draft does discuss both the bootstrapping of identity and the issues of privacy-aware identification. One key aspect is that of bootstrapping a secure conversation between the IoT device and other systems, which includes the challenge of setting-up an encrypted and/or authenticated channel such as those using TLS, Host Identity Protocol (HIP) or Diet HIP. The HIP [68] is a protocol designed to provide a cryptographically secured endpoint to replace the use of IP addresses, which solves a significant problem – IP-address spoofing – in the Internet. Diet HIP [69] is a lighter-weight rendition of the same model designed specifically for IoT and Machine-to-Machine (M2M) interactions. While HIP and Diet HIP solve difficult problems, they have significant disadvantages to adoption. First, they require low-level changes within the IP stack to implement. Second, as they replace traditional IP addressing they require a major change in many existing systems to work. In addition, neither HIP nor Diet HIP address the issues of federated authorisation and delegation.

We proposed [70] using *federated* identity protocols such as OAuth2 [71] with IoT devices, especially around the MQTT protocol [72]. The IOT-OAS [73] work similarly addresses the use of OAuth2 with Common Opensource Application Publishing Platform (CoAP). Other related works include the work of Augusto and Correia [74] have built a secure mobile digital wallet by using OAuth together with eXtensible Messaging and Presence Protocol (XMPP) [75]. In Reference 56, we extended the usage of OAuth2 for IoT devices to include the use of Dynamic Client Registration [76] which allows each device to have its own unique identity, which we discussed as an important point in the section about Cell *A*.

1.2.13   Cell M: access control and device/hardware

There are two challenges to access control at the device level. First, devices are often physically distributed and so an attacker is likely to be able to gain physical access to the device. The challenges here were already discussed in Cell *A*. However, there is a further challenge: access control requires a concept of identity. We cannot restrict or allow access without some form of authentication to the device, and as discussed in our review of Cell *J*, this is a significant challenge. In addition, systems such as Webinos [77] have proposed using policy-based access control mechanisms such as eXtensible Access Control Markup Language (XACML) [78] for IoT devices. However, XACML is relatively heavyweight and expensive to implement [79], especially in the context of low power devices. To address this, Webinos has developed an engine which can calculate the subset of the policy that is relevant to a particular device. Despite this innovation, the storage, transmission and processing costs of XACML are very high for an IoT device.

1.2.14   Cell N: access control and network

There are a number of researchers looking at how to create new lightweight protocols for access control in IoT scenarios. Mahalle *et al.* [80] describe a new protocol for IoT authentication and access control is proposed based on ECC with a lightweight handshake mechanism to provide an effective approach for IoT, especially in mobility cases. Hernández-Ramos *et al.* [81] propose a non-centralised approach for access control that uses ECC once again and supports capability tokens in CoAP.

1.2.15   Cell O: access control and cloud/server

The biggest challenge for privacy is ensuring access control at the server or cloud environment of data collected from the IoT. There is some significant overlap with the area of confidentiality of data in the cloud as well (Cell *C*).

We argued strongly in Reference 70 that existing hierarchical models of access control are not appropriate for the scale and scope of the IoT. There are two main approaches to address this. The first is *policy-based* security models where roles and groups are replaced by more generic policies that capture real-world requirements such as "A doctor may view a patient's record if they are treating that patient in the emergency room". The second approach to support the scale of IoT is user-directed security controls. In Reference 82, a strong case is made for ensuring that users can control access to their own resources and to the data produced by the IoT that relates to those users. The User-Managed Access (UMA) from the Kantara Initiative enhances the OAuth specification to provide a rich environment for users to select their own data sharing preferences [83]. We would argue strongly that this overall concept of user-directed access control to IoT data is one of the most important approaches to ensuring privacy.

Winter [84] argues that contextual approaches must be taken to ensure privacy with the IoT. Many modern security systems use context and reputation to establish trust and to prevent data leaks. Context-based security [85] defines this approach which is now implemented by major Web systems including Google and Facebook.

1.2.16 Cell P: non-repudiation and device/hardware

The biggest challenge in the non-repudiation network with IoT devices is the challenge of using *attestation* for small devices. Attestation is discussed in detail in Cell *D*. Without attestation, we cannot trust that the device system has not been modified and therefore it is not possible to trust any non-repudiation data from the device.

1.2.17 Cell Q: non-repudiation and network

The same challenges apply here as discussed in Cells *B* and *E*. Non-repudiation on the wire requires cryptography techniques and these are often hindered by resource restrictions on small devices. In Reference 86, a non-repudiation protocol for restricted devices is proposed.

1.2.18 Cell R: non-repudiation and cloud/server

This area is unchanged by the IoT, so we do not discuss it any further.

1.2.19 Summary of the review of security issues

In this section, we have proposed a widened ontology for evaluating the security issues surrounding the IoT, and examined the existing literature and research in each of the cells of the expanded matrix. This is an important basis for the next section where we examine the provisions around security and privacy that are available in available middleware for the IoT.

One area that crosses most or all of the cells in our matrix is the need for a holistic and studied approach to enabling privacy in the IoT. As discussed in a number of cells, there are significant challenges to privacy with the increased data and metadata that is being made available by IoT-connected devices. An approach that has been proposed to address this is *Privacy by Design* [87]. This model suggests that systems should be designed from the ground up with the concept of privacy built into the heart of each system. Many systems have added security or privacy controls as "add-ons", with the result that unforeseen attacks can occur.

In reviewing these areas, we identified a list of security properties and capabilities that are important for the security and privacy of IoT. We will use this list in the second part of this chapter as columns in a new table where we evaluate a set of middleware on their provision of these capabilities.

Integrity and confidentiality

The requirement to provide integrity and confidentiality is an important aspect in any network and as discussed in Cells *A–E* there are a number of challenges in this space for IoT.

Access control

Maintaining access control to data that is personal or can be used to extract personal data is a key aspect of privacy. In addition, it is of prime importance with actuators that we do not allow unauthorised access to control aspects of our world.

Policy-based security

Managing security in the scale of IoT is unfeasible in a centralised approach. As we discussed, access control and identity models need to be based on policies such as XACML rather than built in a traditional hierarchical approach.

Authentication

Clearly, in order to respect privacy, IoT systems need a concept of authentication.

Federated identity

As argued in Cell *L*, there is a clear motivation for the use of federated models of identity for authentication in IoT networks.

Attestation

Attestation is an important technique to prevent tampering and hence issues with integrity of data as well as confidentiality in IoT.

Summarisation and filtering

The need to prevent de-anonymisation is a clear driver for systems to provide summarisation and filtering technologies such as stream processing.

Privacy by Design

As discussed above, an important approach to ensuring privacy is to build this into the design of the systems.

Context-based security and reputation

Many modern security models adapt the security based on a number of factors, including location, time of day, previous history of systems, and other aspects known as context. Another related model is that of the reputation of systems, whereby systems that have unusual or less-than-ideal behaviour can be trusted less using probabilistic models. In both cases, there are clear application to IoT privacy as discussed above.

There are of course many other aspects to IoT security and privacy as we have demonstrated in the matrix table and accompanying description of each cell. However, these specific aspects form an effective set of criteria by which to analyse different systems, as we show below.

1.3   Secure middleware for the IoT

1.3.1   Introduction

Middleware has been defined as computer software that has an intermediary function between the various applications of a computer and its operating system [88]. In our case, we are interested in middleware that is specifically designed or adapted to provide capabilities for IoT networks. There are a number of existing surveys of IoT middleware.

Bandyopadhyay *et al.* [89, 90] review a number of middleware systems designed for IoT systems. While they look at security in passing, there is no detailed analysis of the security of each middleware system. Reference 91 calls out the need for security, but no analysis of the approaches or existing capabilities is provided. Reference 92 is a very broad survey paper that addresses IoT middleware loosely.

It is clear then, that a detailed evaluation of security in IoT middleware is a useful contribution to the literature. We therefore identified a set of middleware systems to study.

1.3.2   Review methodology

This set was identified through a combination of the existing literature reviews on IoT middleware [89, 91] together with our own search for middleware systems that explicitly target IoT scenarios. Some of the systems that were included in these papers we excluded from our list on the basis that they were not middleware systems. For example, Reference 91 lists TinyREST [93] as a middleware, but in fact we considered this paper to be the definition of a standard protocol and therefore we excluded it.

Our search strategy was to use a search for the terms ("IoT" OR "Internet of Things") AND "Middleware". We searched only in the subject terms and restricted the search to academic papers written in English. The search was carried out by the Portsmouth University Discovery system which is a metasearch engine. The list of databases that are searched is available in Reference 94. This strategy identified 152 papers. We then manually reviewed the abstracts of these papers to identify a list of functioning middleware systems as opposed to papers that describe other aspects of IoT without describing a middleware system. This produced a list of 22 middleware systems. The table in Figure 1.2 summarises the middleware systems that we reviewed as well as the major findings of the review.

In our study, we looked for the security properties listed in Section 1.2.19. We also identified if the middleware had a clearly defined security model and/or security implementation. In addition, we used some other more general characteristics, including whether the systems supported Simple Object Access Protocol (SOAP)/Web Services (WS), REST, Event-Based models, Semantic approaches, and IoT-specific protocols. Together with the security properties these make up the columns of our summary table.

Middleware	Defined security model	Tangible security architecture	SOAP/WS-*	REST	Event driven	Semantic	IoT-specific protocol support	Integrity and confidentiality	Access control	User-centric access control	Policy-based security	Authentication	Federated identity	Attestation	Summarisation and filtering	Privacy by design	Context-based security/reputation
ASPIRE	N	N	Y	N	N	N	N	–	–	–	–	–	–	–	–	–	–
CBCPM	N	N	N	Y	Y	N	N	–	–	–	–	–	–	–	–	–	–
Dioptase	N	N	N	N	Y	Y	N	–	–	–	–	–	–	–	Y	–	–
DREMS	Y	Y	N	N	Y	N	Y	Y	Y	N	N	Y	N	N	N	N	N
EDSOA	N	N	Y	N	Y	N	N	–	–	–	–	–	–	–	Y	–	–
GSN	N	N	N	N	Y	N	N	–	–	–	–	–	–	–	Y	–	–
Hydra/LinkSmart	Y	Y	Y	N	N	Y	N	Y	Y	N	N	Y	N	N	N	N	N
ISMB/VIRTUS	Y	Y	N	N	Y	N	N	Y	Y	N	N	Y	Y	N	N	N	N
MOSDEN	N	N	N	N	N	N	N	–	–	–	–	–	–	–	Y	–	–
NAPS	Y	N	–	–	Y	Y	Y	–	–	–	–	–	–	–	–	–	–
OpenIoT	N	N	–	–	–	N	N	–	–	–	–	–	–	–	Y	–	–
SBIOTCM	N	N	Y	N	N	N	N	–	N	N	N	N	N	N	N	N	N
SIRENA	Y	Y	Y	N	N	Y	N	Y	Y	N	N	Y	N	N	N	N	N
SMEPP	Y	Y	Y	N	Y	N	N	Y	Y	N	N	Y	N	N	N	N	N
SOCRADES	Y	Y	Y	N	N	Y	–	–	N	N	N	N	N	N	N	N	N
Thingsonomy	N	N	–	–	Y	Y	N	–	–	–	–	–	–	–	–	–	–
UBIROAD	N	N	Y	N	N	Y	N	–	–	–	–	–	–	–	N	N	N
UBISOAP	N	N	Y	N	N	N	N	–	–	–	–	–	–	–	N	–	–
UBIWARE	Y	N	N	N	N	Y	N	–	–	–	Y	–	–	–	–	–	–
WEBINOS	Y	Y	Y	Y	N	N	N	Y	Y	N	Y	Y	Y	Y	N	N	N
WHEREX	N	N	–	–	Y	–	–	–	–	–	–	–	–	–	–	–	–
XMPP	Y	Y	N	N	Y	N	N	Y	Y	N	N	Y	Y	N	N	N	N

Figure 1.2 Summary of reviewed middleware systems and major properties

Below are the specific details of each middleware system.

1.3.3 ASPIRE

ASPIRE Project (Advanced Sensors and lightweight Programmable middleware for Innovative Rfid Enterprise applications) [95] is a European Union (EU)-funded project that created an open, royalty-free middleware for RFID-based applications. There is insufficient description of the security architecture to make any sensible review.

1.3.4 UBIWARE

The UBIWARE project is a smart semantic middleware for Ubiquitous Computing [96]. The security model for UBIWARE is not clearly described in the original paper, but an additional paper describes a model called Smart Ubiquitous Resource Privacy and Security (SURPAS) [97], which provides a security model for UBIWARE. UBIWARE is designed to utilise the semantic Web constructs, and SURPAS utilises the same model of semantic Web as the basis for the abstract and concrete security architectures that it proposes. The model is highly driven by policies and these can be stored and managed by external parties. In particular, the SURPAS architecture is highly dynamic, allowing devices to take on board new roles or functions at run-time. While the SURPAS model describes a theoretical solution to the approach, there are few details on the concrete instantiation. For example, while the model defines a policy-based approach to access control, there are no clearly defined policy languages chosen. There is no clear model of identity or federation, and there is no clear guidance on how to ensure that federated policies that are stored on external servers are protected and maintain integrity. The model does not address any edge computing approaches or filtering/summarisation of IoT data. However, the overall approach of using ontologies and basing policies on those ontologies is very powerful.

1.3.5 UBIROAD

The UBIROAD middleware [98] is a specialisation of the UBIWARE project specifically targeting traffic, road management, transport management, and related use-cases. There is insufficient description of the security and trust architecture to make any meaningful review.

1.3.6 UBISOAP

ubiSOAP [99] is a Service-Oriented Architecture (SOA) approach that builds a middleware for Ubiquitous Computing and IoT based on the WS standards and SOAP. There is insufficient description of the security and trust architecture to make any meaningful review.

1.3.7 SMEPP

Secure Middleware for P2P (SMEPP) [100] is an IoT middleware explicitly designed to be secure, especially dealing with challenges in the peer-to-peer model. SMEPP security is based around the concept of a group. When a peer attempts to join a group, the system relies on challenge–response security to implement mutual authentication. At this point, the newly joined peer is issued a shared session key which is shared by all members of the group. SMEPP utilises elliptic key cryptography to reduce the burden of the security encryption onto smaller devices. Overall SMEPP has addressed security effectively for peer-to-peer groups, but assumes a wider PKI for managing the key model used within each group. In addition, there is no discussion of access control or federated identity models, which are important for IoT scenarios. The model is that any member of the group can read data published to the group using the shared session key.

1.3.8 SOCRADES

SOCRADES [101] is a middleware specifically designed for manufacturing shop floors and other industrial environments. Based on SOAP and the WS stack, it utilises the security models of the WS stack, in particular the WS-Security standard for encryption and message integrity. There is no special support for federation, tokens or policy-based access control (instead relying on role-based access control). The resulting eXtensible Markup Language (XML) approach is very heavyweight for IoT devices and costly in terms of network and power [102]. In addition, the lack of explicit support for tokens and federated security and identity models creates a significant challenge in key distributions and centralised identity for this approach.

1.3.9 SIRENA

SIRENA (Service Infrastructure for Real-time Embedded Networked Devices) [103] is a SOAP/WS-based middleware for IoT and embedded devices. While there is little description of the security framework in SIRENA, it does show the use of the WS-Security specification. As previously discussed, this approach is very heavyweight, has issues with key distribution, federated identity, and access control.

1.3.10 WHEREX

WhereX [104] is an event-based middleware for the IoT. There is insufficient description of the security and trust architecture to make any meaningful review.

1.3.11 WEBINOS

The Webinos [77] system has a well-thought through security architecture. The Webinos system is based around the core concept of devices being in the personal control of users and therefore having each user having a "personal zone" to protect. This is a more advanced concept but in the same vein as the protected sub-domains in VIRTUS. In the Webinos model, each user has a cloud instance – known as the Personal Zone Hub (PZH) that supports their devices. The PZH acts as a service to collect and offer access to data and capabilities of the user's devices. The PZH acts as a CA, issuing certificates to the devices that are used for mutual authentication using TLS. Users authenticate to the PZH using the OpenID protocol. On the device, a communications module known as the Personal Zone Proxy (PZP) handles all communications with the PZH.

The idea of the Personal Zone may have significant issues however, when a single device is used by many different people (e.g. the in-car system in a taxi as opposed to a personal vehicle). These issues are not addressed in Webinos, though they are called out in the lessons learnt.

Webinos utilises policy-based access control modelled in XACML [78]. The system pushes XACML policies out to devices to limit the spread of personal and contextual data.

Webinos addresses the issue of software modification using an attestation Application Programming Interface (API), which can report whether the software running is the correct level. This requires the device to be utilising TPM hardware that can return attestation data.

Webinos also addresses the issue of using secure storage on devices where the device has such storage.

While the Webinos project does address many of the privacy concerns of users through the use of the PZH, there is clearly further work that could be done. In particular, the ability for users to define what data they share with other users or other systems using a protocol such as OAuth2 [71], and the ability to install filters or other anonymising or data reduction aggregators into the PZH are lacking. One other aspect of Webinos that is worth drawing attention to is the reliance on a certain size of device: the PZP that is needed on the device is based on the *node.js* framework and therefore the device needs to be of a certain size (e.g. a 32-bit processor running a Linux derivative or similar) to participate in Webinos.

1.3.12   GSN

The GSN (Global Sensor Networks) framework [105] defines a middleware for the IoT that requires little or no programming. The security architecture of the system is not described in any detail; there are diagrams of the container architecture which point to access control and integrity checks, but unfortunately there is not sufficient discussion to be able to categorise or evaluate the approach taken.

1.3.13   MOSDEN

MOSDEN (Mobile Sensor Data Processing Engine) [106] is an extension of the GSN approach (see above) which is explicitly targeted at *opportunistic* sensing from restricted devices. As with GSN, there is insufficient description of the security and trust architecture to make any meaningful review.

1.3.14   Thingsonomy

Thingsonomy [107] is an event-based publish–subscribe-based approach that applies semantic technology and semantic matching to the events published within the system. There is no description of a security model.

1.3.15   OpenIoT

OpenIoT is an open cloud-based middleware for the IoT. It also extends the GSN framework. There is insufficient description of the security model to make any meaningful review.

1.3.16   Dioptase

Dioptase [108] is a RESTful stream-processing middleware for IoT. Dioptase does address a number of useful aspects for privacy, including intermediate stream processing of data, summarisation, and filtering. However, there is no detailed security architecture or description and the security model is left as an item of future work.

1.3.17   VIRTUS

The VIRTUS middleware [109] utilises the core security features of the XMPP to ensure security. This includes tunnelling communications over TLS, authentication via Simple Authentication and Security Layer (SASL), and access control via XMPP's built-in mechanisms. SASL is a flexible mechanism for authentication which supports a number of different systems including token-based approaches such as OAuth2 or Kerberos, username/password, or X.509 certificates. For client-to-server-based communications, it is not clear from the description which of these methods are actually implemented within VIRTUS. For server-to-server communications, there is specified use of SASL to ensure full server federation.

While the VIRTUS model does not describe the challenges of implementing a personal instance of middleware for single users or devices, there is a concept of edge computing described, where some interactions may happen within an edge domain (e.g. within a house) and lower security is required within that domain while higher security is expected when sharing that data outside. This model is fairly briefly described but provides an interesting approach. One challenge is that there are multiple assumptions to this: first, that security within the limited domain needs less security, when there may be attackers within that perimeter. Second, that the open channel to the wider Internet cannot be misused to attack the edge network. The ability to calculate, summarise, and/or filter data from the edge network before sharing it is also not discussed except in very granular terms (e.g. some data are available, other data are not).

1.3.18   Hydra/LinkSmart

Hydra [110] was a EU-funded project which has since been extended and renamed as LinkSmart. The Hydra team published a detailed theoretical model of a policy-based security approach [111]. This model is based on using lattices to define the flow of information through a system. This model provides a language-based approach to security modelling. However, while this paper is published as part of the Hydra funded project, there is no clear implementation of this in the context of IoT or description of how this work can benefit the IoT world. However, because Hydra/LinkSmart is an Open-Source project [112] with documentation beyond the scientific papers, it is possible to understand the security model in greater detail by review of this project.

The Hydra and LinkSmart architectures are both based on the WS specifications, building on the SOAP [113], which in turn builds on XML [114]. The security model is described in some detail in the LinkSmart documentation [115]. The model utilises XML Security [116]. There are significant challenges in using this model in the IoT world. XML Security has a number of performance issues which are exacerbated by the need to utilise this in an IoT context. For example, any digital signature in XML Security needs a process known as XML Canonicalisation (XML C14N). XML Canonicalisation is a costly process in both time and memory. Binna [117] shows that the memory usage is more than $10\times$ the size of the message in memory (and XML messages are already large for IoT devices). The Hydra/LinkSmart approach also uses symmetric keys for security which is a challenge for IoT because each key must be uniquely created, distributed, and updated upon expiry into each device creating a major key management issue.

Hydra/LinkSmart offers a service called the TrustManager. This is a system that uses the cryptographic capabilities to support a trusted identity for IoT devices. This works with a PKI and certificates to ensure trust. Once again there are challenges in the distribution and management of the certificates to the devices which are not addressed in this middleware.

The Hydra middleware does not offer any policy-based access control for IoT data, and does not address the secure storage of data for users, nor offer any user-controlled models of access control to user's data.

1.3.19 EDSOA

An Event-Driven Service-Oriented Architecture (EDSOA) for the Internet of Things Service Execution [118] describes an approach that utilises an event-driven SOA. There is no security model described.

1.3.20 DREMS

Distributed Real-Time Managed Systems (DREMS) [119] is a combination of software tooling and a middleware runtime for IoT. It includes Linux Operating System extensions as well. DREMS is based on an actor [120] model has a well-defined security model that extends to the operating system. The security model includes the concept of multi-level security (MLS) for communications between a device and the actor. The MLS model is based on *labelled* communications. This ensures that data can only flow to systems that have a higher *clearance* than the data being transmitted. This is a very powerful security model for government and military use-cases. However, this approach does not address needs-based access control. For example, someone with *Top Secret* clearance may read data that is categorised as *Secret* even if they have no business reason to utilise that data. The weaknesses of this model have been shown with situations such as the Snowden revelations.

1.3.21   XMPP

Iivari *et al.* [121] describe how the XMPP architecture can be applied to the challenges of M2M and hence the IoT, together with a proof-of-concept approach. The system relies on the set of XMPP extensions around publish–subscribe and the related XMPP security models to implement security. This includes TLS for encryption, and access control models around publish–subscribe. There is also a discussion about leakage of information such as *presence* from devices. The proof-of-concept model did not include any federated identity models, but did utilise a One-Time Password (OTP) model on top of XMPP to address the concepts such as temporary loans of devices.

1.3.22   Cloud-based car parking middleware

In Reference 122, the authors describe an Open Services Gateway initiative (OSGi)-based middleware for smart cities enabling IoT-based car parking. There is no description of the security architecture.

1.3.23   NAPS

The *Naming, Addressing, and Profile Server* (NAPS) [123] describes a heterogeneous middleware for IoT based on unifying data streams from multiple IoT approaches. Based on RESTful APIs, the NAPS approach includes a key component handling Authentication, Authorisation, and Accounting (AAA). The design is based on the Network Security Capability model defined in the ETSI M2M architecture [124]. However, the main details of the security architecture have not yet been implemented and have been left for future work. There is no consideration of federated identity or policy-based access control.

1.3.24   SBIOTCM

In *A SOA Based IOT Communication Middleware* [125] is a middleware based on SOAP and WS. There is no security model described.

1.4   Summary of IoT middleware security

In reviewing both the security and privacy challenges of the wider IoT and a structured review of more than 20 middleware platforms, we have identified some key categories that can be applied across these areas.

First, we must deal with the significant proportion of the systems that did not address security, left it for further work, or did not describe the security approach in any meaningful detail. This category includes WHEREX, ASPIRE, GSN, Thingsonomy, Dioptase, OpenIoT, UBIROAD, UBISOAP, CBCPM, and EDSOA.

There are two further approaches (UBIWARE and NAPS) that offer theoretical models but did not demonstrate any real-world implementation or concrete approach.

The next clear category is those middlewares that apply the SOAP/WS model of security. This includes SOCRADES, SIRENA, and Hydra/LinkSmart. As we have discussed in the previous sections, there are significant challenges in performance, memory footprint, processor power, and usability of these approaches when used with the IoT.

Two of the approaches delegate the model to the XMPP standards: VIRTUS and XMPP [121, 126]. XMPP also has the complexity of XML, but avoids the major performance overheads by using TLS instead of XML Encryption and XML Security.

This finally leaves a few unique approaches, each of which brings their own unique benefits.

DREMS is the only system to provide MLS based on the concept of security clearances. While this model is attractive to government and military circles (because of the classification systems used in those circles), we would argue that it fails in many regards for IoT. In particular, there are no personal controls, no concept of federated identity, and no policy-based access controls in this model.

SMEPP offers a model based on PKIs and shared session keys. We would argue this approach has a number of challenges scaling to the requirements of the IoT. First, there are significant issues in key distribution and key revocation. Second, this model creates a new form of perimeter based on the concept of a shared session key. That means that if one device is compromised then the data and control of all the devices in that group are also compromised.

Only Dioptase supports the concept of stream processing in the cloud, which we argue is a serious requirement for the IoT. The requirement is to be able to filter, summarise, and process streams of data from devices to support anonymisation and reduction of data leakage.

Finally, we identified that the most advanced approach is that proposed by Webinos. Webinos utilises some key technologies to provide a security and privacy model. First, this uses policy-based access control (XACML). The model does not however support user-guided access control mechanisms such as OAuth2 or UMA.

Webinos does support the use of federated identity tokens (OpenID), but only from users to the cloud, as opposed to devices to the cloud. We and others have proposed the model of using federated identity tokens from the device to the cloud in References 56, 70, 73.

The contribution of the Webinos work with the largest potential impact is the concept of PZH, which is a cloud service dedicated to a single user to handle the security and privacy requirements of that user. There is, however, further research around this area: the PZH model from Webinos does not examine many of the challenges of how to implement the PZH in real life. For example, user registration, cloud hosting, and many other aspects need to be defined in more detail before the Webinos PZH model is practicable for real-world projects. In addition, there are challenges using the PZH model with smaller devices because of the requirement to use the PZP.

1.4.1   Overall gaps in the middleware

When we look at the requirements for security and privacy of the IoT, we can see there are some gaps that are not provided by any of the reviewed middleware systems.

- None of the middleware systems explicitly applied the concept of Privacy by Design in designing a middleware directly to support privacy, although Webinos did exhibit many of the characteristics of a system that used this approach.

- None of the models applied any concepts of context-based security or reputation to IoT devices.
- None of the middleware systems offered a user-centric model of access control.
- None of the middleware systems utilised federated identity at the device level.

1.5   Discussion

1.5.1   Contributions

In this chapter, we have taken a two-phase approach to reviewing the available literature around the security and privacy of IoT devices.

In the first part, we created a matrix of security challenges that applied the existing CIA+ model to three distinct areas: device, network, and cloud. This new model forms a clear contribution to the literature. In each of these areas, we either identified that there a no distinct challenges or we identified a set of clear challenges that is either unique to or exacerbated by the IoT.

In the second part, we used a structured search approach to identify 22 specific IoT middleware frameworks and we analysed the security models of each of those. While there are existing surveys of IoT middleware, none of them focussed on a detailed analysis of the security of the surveyed systems and therefore this has a clear contribution to the literature.

1.5.2   Further work

In our survey, we have identified some clear gaps. Over half the surveyed systems had either no security or no substantive discussion of security. Out of 22 surveyed systems, we found very few that addressed a significant proportion of the major challenges that we identified in the first section. We found certain aspects that were identified in the first section that were not addressed by any of the surveyed systems. Based on this, we believe there is a significant opportunity to contribute to the research by creating a middleware for IoT that addresses these gaps.

- To define a model and architecture for IoT middleware that is designed from the start to enable privacy and security (Privacy by Design).
- Second, to bring together the best practice into a single middleware that includes: federated identity (for users and devices), policy-based access control, UMA to data, stream processing in the cloud.
- Third, there is considerable work to be done to define a better model around the implementation challenges for the concept of a personal cloud service (e.g. the Webinos PZH). This includes the hosting model, bootstrapping, discovery, and usage for smaller devices.
- Finally, creating a middleware system that applies context-based security and reputation to IoT middleware.

References

[1] K. Ashton. "That 'internet of things' thing", *RFiD Journal*, **22**, pp. 97–114 (2009).

[2] Cluster CERP-IoT. "Internet of things, strategic research roadmap", *European Commission* (2009) [Online]. Available: http://www.grifs-project. eu/data/ File/CERP-IoT%20SRA_IoT_v11.pdf. (Visited on 04/04/2016).

[3] D. Evans. "The internet of things", *How the Next Evolution of the Internet is Changing Everything, Whitepaper, Cisco Internet Business Solutions Group (IBSG)* (2011).

[4] C. P. Pfleeger, S. L. Pfleeger. *Security in Computing* (Prentice Hall Professional Technical Reference, Indianapolis, IN, 2002).

[5] A. Simmonds, P. Sandilands, L. Van Ekert. "An ontology for network security attacks", *Applied Computing*, pp. 317–323 (Springer, Berlin, 2004).

[6] S. B. Furber. *ARM system Architecture* (Addison-Wesley Longman Publishing Co., Inc., Boston, MA, 1996).

[7] M. A. M. Vieira, C. N. Coelho Jr, D. C. da Silva Jr, J. M. da Mata. "Survey on wireless sensor network devices", *Emerging Technologies and Factory Automation, 2003. Proceedings. ETFA'03. IEEE Conference*, volume 1, pp. 537–544 (IEEE, Piscataway, NJ, 2003).

[8] Arduino. "Arduino" [Online]. Available: http://arduino.cc/ (2015) (Visited on 04/04/2016).

[9] R. Chakravorty, J. Cartwright, I. Pratt. "Practical experience with TCP over GPRS", *Global Telecommunications Conference, 2002. GLOBECOM'02. IEEE*, volume 2, pp. 1678–1682 (IEEE, Piscataway, NJ, 2002).

[10] S. P. Skorobogatov. "Semi-invasive attacks: a new approach to hardware security analysis", University of Cambridge, Technical Report 630 (2005).

[11] S. Y. Yan. "Side-channel attacks", *Cryptanalytic Attacks on RSA*, pp. 207–222 (Springer, Berlin, 2008).

[12] V. Lomne, A. Dehaboui, P. Maurine, L. Torres, M. Robert. "Side channel attacks", *Security Trends for FPGAS*, pp. 47–72 (Springer, Berlin, 2011).

[13] R. Watro, D. Kong, S.-f. Cuti, C. Gardiner, C. Lynn, P. Kruus. "Tinypk: securing sensor networks with public key technology", *Proceedings of the Second ACM Workshop on Security of Ad Hoc and Sensor Networks*, pp. 59–64 (ACM, New York, NY, 2004).

[14] S. Yi, R. Kravets. "Key management for heterogeneous ad hoc wireless networks", *Network Protocols, 2002. Proceedings of the 10th IEEE International Conference on*, pp. 202–203 (IEEE, Piscataway, NJ, 2002).

[15] P. McDaniel, S. McLaughlin. "Security and privacy challenges in the smart grid", *Security & Privacy, IEEE*, **7(3)**, pp. 75–77 (2009).

[16] H. Khurana, M. Hadley, N. Lu, D. A. Frincke. "Smart-grid security issues", *Security & Privacy, IEEE*, **8(1)**, pp. 81–85 (2010).

[17] K. Hill. "When 'smart homes' get hacked: I haunted a complete stranger's house via the internet – Forbes" [Online]. Available: http://www.forbes.com/ sites/kashmirhill/2013/07/26/smart-homes-hack/ (2013) (Visited on 07/09/ 2015).

[18]   H. Bojinov, Y. Michalevsky, G. Nakibly, D. Boneh. "Mobile device identification via sensor fingerprinting", *The Computing Research Repository (CoRR)*, vol. abs/1408.1416 (2014).

[19]   N. Koblitz. "Elliptic curve cryptosystems", *Mathematics of Computation*, **48(177)**, pp. 203–209 (1987).

[20]   V. S. Miller. "Use of elliptic curves in cryptography", *Advances in Cryptology-CRYPTO'85 Proceedings*, pp. 417–426 (Springer, Berlin, 1986).

[21]   R. L. Rivest, A. Shamir, L. Adleman. "A method for obtaining digital signatures and public-key cryptosystems", *Communications of the ACM*, **21(2)**, pp. 120–126 (1978).

[22]   N. Gura, A. Patel, A. Wander, H. Eberle, S. Shantz. "Comparing elliptic curve cryptography and RSA on 8-bit CPUs", M. Joye, J.-J. Quisquater (Eds.), *Cryptographic Hardware and Embedded Systems – CHES 2004*, volume 3156 of *Lecture Notes in Computer Science*, pp. 119–132 (Springer, Berlin, 2004) [Online]. Available: http://dx.doi.org/10.1007/978-3-540-28632-5_9. (Visited on 04/04/2016).

[23]   M. Sethi. "Security in smart object networks", Master's thesis, Aalto University School of Science, Espoo, Finland (2012).

[24]   M. Sethi, J. Arkko, A. Keranen. "End-to-end security for sleepy smart object networks", *Local Computer Networks Workshops (LCN Workshops), 2012 IEEE 37th Conference on*, pp. 964–972 (IEEE, Piscataway, NJ, 2012).

[25]   D. Landman. "DavyLandman/AESlib" [Online]. Available: https://github.com/DavyLandman/AESLib (2015) (Visited on 07/09/2015).

[26]   T. Dierks and E. Rescorla. "The transport layer security (TLS) protocol: version 1.2". IETF RFC 5246 (2008).

[27]   M. Koschuch, M. Hudler, M. Krüger. "Performance evaluation of the TLS handshake in the context of embedded devices", *Data Communication Networking (DCNET), Proceedings of the 2010 International Conference on*, pp. 1–10 (IEEE, Piscataway, NJ, 2010).

[28]   V. Perelman, M. Ersue. "TLS with PSK for constrained devices", available from http://www.lix.polytechnique.fr/hipercom/SmartObjectSecurity/papers/VladislavPerelman.pdf. Last accessed 14 June 2016 (2012).

[29]   E. Rescorla, N. Modadugu. "Datagram transport layer security", Technical report (2006).

[30]   S. Keoh, S. Kumar, O. Garcia-Morchon. "Securing the IP-based internet of things with DTLS", *Working Draft, February* (2013).

[31]   F. Audet, C. Jennings. "Network address translation (NAT) behavioral requirements for unicast UDP", Technical report (2007).

[32]   H. Tschofenig, T. Fossati. "A TLS/DTLS 1.2 profile for the internet of things", available from https://tools.ietf.org/html/draft-ietf-dice-profile-17. Last accessed 14 June 2016 (2015).

[33]   N. Cam-Winget, R. Housley, D. Wagner, J. Walker. "Security flaws in 802.11 data link protocols", *Communications of the ACM*, **46(5)**, pp. 35–39 (2003).

[34]   M. Ryan. "Bluetooth: with low energy comes low security", *WOOT* (2013).

[35]   B. Botezatu. "25 percent of wireless networks are highly vulnerable to hacking attacks, Wi-fi security survey reveals – hotforsecurity" [Online]. Available: http://www.hotforsecurity.com/blog/25-percent-of-wireless-networks-are-highly-vulnerable-to-hacking-attacks-wi-fi-security-survey-reveals-1174.html (2011) (Visited on 07/14/2015).

[36]   S. Radomirovic. "Towards a model for security and privacy in the internet of things", *Proceedings of the First International Workshop on Security of the Internet of Things*, Tokyo, Japan (2010).

[37]   Fitbit. "Fitbit official site for activity trackers and more" [Online]. Available: http://www.fitbit.com/ (2015) (Visited on 07/09/2015).

[38]   Zee. "Fitbit users are unwittingly sharing details of their sex lives with the world" (2011) [Online]. Available: http://thenextweb.com/insider/2011/07/03/fitbit-users-are-inadvertently-sharing-details-of-their-sex-lives-with-the- world/, (Visited on 06/04/2013).

[39]   A. Rendle. "Who owns the data in the internet of things?" [Online]. Available: http://www.taylorwessing.com/download/article_data_lot.html (2014) (Visited on 06/08/2015).

[40]   C. Murphy. "Internet of things: who gets the data? – information-week" [Online]. Available: http://www.informationweek.com/strategic-cio/executive-insights-and-innovation/internet-of-things-who-gets-the-data/a/d-id/1252701 (2014) (Visited on 06/08/2015).

[41]   A. Narayanan, V. Shmatikov. "Robust de-anonymization of large sparse datasets", *Security and Privacy, 2008. SP 2008. IEEE Symposium on,* pp. 111–125 (IEEE, Piscataway, NJ, 2008).

[42]   J. Card. "Anonymity is the Internet's next big battleground" [Online]. Available: http://www.theguardian.com/media-network/2015/jun/22/anonymity-internet-battleground-data-advertisers-marketers (2015) (Visited on 07/ 13/2015).

[43]   J. Larson, N. Perlroth, S. Shane. "The NSA's secret campaign to crack, undermine internet encryption – propublica" [Online]. Available: http://www.propublica.org/article/the-nsas-secret-campaign-to-crack-undermine-internet-encryption (2013) (Visited on 06/08/2015).

[44]   L. Levinson. "Secrets, lies and Snowden's email: why I was forced to shut down lavabit" [Online]. Available: http://www.theguardian.com/commentis free/2014/may/20/why-did-lavabit-shut-down-snowden-email   (2014) (Visited on 06/08/2015).

[45]   J. Ball. "NSA stores metadata of millions of web users for up to a year, secret files show" [Online]. Available: http://www.theguardian.com/world/2013/sep/30/nsa-americans-metadata-year-documents (2013) (Visited on 06/08/2015).

[46]   D. Goodin. "New Linux worm targets routers, cameras, internet of things devices", available from http://arstechnica.com/security/2013/11/new-linux-worm-targets-routers-cameras-internet-of-things-devices/, last accessed 14 June 2016 (2013).

[47]   A.-R. Sadeghi, C. Stüble. "Property-based attestation for computing platforms: caring about properties, not mechanisms", *Proceedings of the 2004 Workshop on New Security Paradigms*, pp. 67–77 (ACM, New York, NY, 2004).

[48]   E. Brickell, J. Camenisch, L. Chen. "Direct anonymous attestation", *Proceedings of the 11th ACM Conference on Computer and Communications Security*, pp. 132–145 (ACM, New York, NY, 2004).

[49]   A. Seshadri, A. Perrig, L. Van Doorn, P. Khosla. "SWATT: software-based attestation for embedded devices", *Security and Privacy, 2004. Proceedings. 2004 IEEE Symposium on*, pp. 272–282 (IEEE, Piscataway, NJ, 2004).

[50]   H. M. Deitel. *An Introduction to Operating Systems*, volume 3 (Addison-Wesley, Reading, MA, 1984).

[51]   TCG. "Trusted computing group – home" [Online]. Available: http://www. trustedcomputinggroup.org/ (2015) (Visited on 06/08/2015).

[52]   Samsung. "Mobile Enterprise Security – Samsung KNOX" [Online]. Available: https://www.samsungknox.com/en (2015) (Visited on 03/24/2015).

[53]   T. Morris. "Trusted platform module" *In Encyclopedia of Cryptography and Security*, pp. 1332–1335 (2011).

[54]   A. Paverd, A. Martin. "Hardware security for device authentication in the smart grid", *First Open EIT ICT Labs Workshop on Smart Grid Security – SmartGridSec12* (Berlin, Germany, 2012) [Online]. Available: http://link.springer.com/chapter/10.1007/978-3-642-38030-3_5. (Visited on 04/04/2016).

[55]   K. Andersson, P. Szewczyk. "Insecurity by obscurity continues: are ADSL router manuals putting end-users at risk", *In Proceedings of the 9th Australian Information Security Management Conference*, available from http://ro.ecu.edu.au/cgi/viewcontent.cgi?article=1106&context=ism   Last accessed 14 June 2016 (2011).

[56]   P. Fremantle, J. Kopeckỳ, B. Aziz. "Web API management meets the internet of things", *Services and Applications over Linked APIs and Data – SALAD2015* (2015) vol. 1359, pp. 1–9.

[57]   J. Newsome, E. Shi, D. Song, A. Perrig. "The Sybil attack in sensor networks: analysis and defenses", *Proceedings of the Third International Symposium on Information Processing in Sensor Networks*, pp. 259–268 (ACM, New York, NY, 2004).

[58]   A. Perrig, R. Szewczyk, J. Tygar, V. Wen, D. E. Culler. "Spins: security protocols for sensor networks", *Wireless Networks*, **8(5)**, pp. 521–534 (2002).

[59]   A. Jøsang, R. Ismail, C. Boyd. "A survey of trust and reputation systems for online service provision", *Decision Support Systems*, **43(2)**, pp. 618–644 (March 2007).

[60]   A. Abdul-Rahman, S. Hailes. "Supporting trust in virtual communities", *HICSS'00: Proceedings of the 33rd Hawaii International Conference on System Sciences – Volume 6* (IEEE Computer Society, Washington, DC, 2000).

[61]   K. Fullam, K. Barber. "Learning trust strategies in reputation exchange networks", *AAMAS '06: Proceedings of the Fifth International Joint Conference on Autonomous Agents and Multiagent Systems*, pp. 1241–1248 (ACM Press, New York, NY, 2006).

[62]   G. C. Silaghi, A. Arenas, L. M. Silva. "Reputation-based trust management systems and their applicability to grids", Technical Report TR-0064, Institutes on Knowledge and Data Management and System Architecture, CoreGRID – Network of Excellence (February 2007).

[63]   C. Zouridaki, B. L. Mark, M. Hejmo, R. K. Thomas. "Hermes: a quantitative trust establishment framework for reliable data packet delivery in MANETs", *Journal of Computer Security*, **15(1)**, pp. 3–38 (2007).

[64]   C. Zouridaki, B. L. Mark, M. Hejmo, R. K. Thomas. "E-hermes: a robust cooperative trust establishment scheme for mobile ad hoc networks", *Ad Hoc Networks*, **7(6)**, pp. 1156–1168 (2009).

[65]   D. Chen, G. Chang, D. Sun, J. Li, J. Jia, X. Wang. "TRM-IoT: a trust management model based on fuzzy reputation for internet of things", *Computer Science and Information Systems*, **8(4)**, pp. 1207–1228 (2011).

[66]   P. Michiardi, R. Molva. "Core: a collaborative reputation mechanism to enforce node cooperation in mobile ad hoc networks", *Advanced Communications and Multimedia Security*, pp. 107–121 (Springer, Berlin, 2002).

[67]   O. Garcia-Morchon. "Security considerations in the IP-based Internet of Things", Internet Draft, IETF (September 2013) [Online]. Available: http://tools.ietf.org/html/draft-garcia-core-security-06. (Visited on 04/04/2016).

[68]   R. Moskowitz and P. Nikander. "Host identity protocol architecture", IETF RFC 4423 (2006).

[69]   R. Moskowitz and R. Hummen. "Hip diet exchange (DEX)", IETF Internet Draft available from https://datatracker.ietf.org/doc/draft-moskowitz-hip-dex/. Last accessed 14 June 2016 (2016).

[70]   P. Fremantle, B. Aziz, P. Scott, J. Kopecky. "Federated identity and access management for the Internet of Things", *Third International Workshop on the Secure IoT*, Wroclaw, Poland (2014).

[71]   D. Hammer-Lahav, D. Hardt. "The OAuth2.0 authorization protocol. 2011", Technical report, IETF Internet Draft (2011).

[72]   D. Locke. "MQ telemetry transport (MQTT) v3.1 protocol specification", *IBM Developer Works Technical Library]* [Online]. Available: http://www.ibm.com/developerworks/webservices/library/ws-mqtt/index.html   (2010) (Visited on 04/04/2016).

[73]   S. Cirani, M. Picone, P. Gonizzi, L. Veltri, G. Ferrari. "IoT-OAS: an OAuth-based authorization service architecture for secure services in IoT scenarios" in IEEE Sensors Journal, vol. 15, no. 2, pp. 1224–1234, (2015).

[74]   A. B. Augusto, M. E. Correia. "An XMPP messaging infrastructure for a mobile held security identity wallet of personal and private dynamic identity attributes", *Ninth national conference on XML, its Associated Technologies and its Applications (XATA'2011)*, Vila do Conde, Portugal (2011).

[75]   P. Saint-Andre. "Extensible messaging and presence protocol (XMPP): core", IETF RFC 6120 (2011).

[76]   N. Sakimura, J. Bradley, M. Jones. "OpenID connect dynamic client registration 1.0 incorporating errata set 1" (2014).

[77]   H. Desruelle, J. Lyle, S. Isenberg, F. Gielen. "On the challenges of building a web-based ubiquitous application platform", *Proceedings of the 2012 ACM Conference on Ubiquitous Computing*, pp. 733–736 (ACM, New York, NY, 2012).

[78]   S. Godik, A. Anderson, B. Parducci, P. Humenn, S. Vajjhala. "Oasis extensible access control 2 markup language (XACML) 3", Technical report, OASIS (2002).

[79]   F. Turkmen, B. Crispo. "Performance evaluation of XACML PDP implementations", *Proceedings of the 2008 ACM Workshop on Secure Web Services*, pp. 37–44 (ACM, New York, NY, 2008).

[80]   P. N. Mahalle, B. Anggorojati, N. R. Prasad, R. Prasad. "Identity establishment and capability based access control (IECAC) scheme for Internet of Things", *Wireless Personal Multimedia Communications (WPMC), 2012 15th International Symposium on*, pp. 187–191 (IEEE, Piscataway, NJ, 2012).

[81]   J. L. Hernández-Ramos, A. J. Jara, L. Marın, A. F. Skarmeta. "Distributed capability-based access control for the internet of things", *Journal of Internet Services and Information Security (JISIS)*, **3(3/4)**, pp. 1–16 (2013).

[82]   H. Tschofenig, E. Maler, E. Wahlstroem and S. Erdtman. "Authentication and authorization for constrained environments using OAuth and UMA", IETF Internet-Draft, available from https://tools.ietf.org/html/draft-maler-ace-oauth-uma-00. Last accessed 14 June 2016 (2015).

[83]   Kantara Initiative. "User managed access (UMA) V1.0.1 Specifications" (2016).

[84]   J. S. Winter. "Privacy and the emerging internet of things: using the framework of contextual integrity to inform policy", *Pacific Telecommunication Council Conference Proceedings 2012*, Honolulu, Hawaii (2012).

[85]   R. Montanari, A. Toninelli, J. M. Bradshaw. "Context-based security management for multi-agent systems", *Multi-Agent Security and Survivability, 2005 IEEE Second Symposium on*, pp. 75–84 (IEEE, Piscataway, NJ, 2005).

[86]   K.-W. Park, H. Seok, K.-H. Park. "PKASSO: towards seamless authentication providing non-repudiation on resource-constrained devices", *Advanced Information Networking and Applications Workshops, 2007, AINAW'07. 21st International Conference on*, volume 2, pp. 105–112 (IEEE, Piscataway, NJ, 2007).

[87]   A. Cavoukian. "Privacy in the clouds", *Identity in the Information Society*, **1(1)**, pp. 89–108 (2008).

[88]   P. Hanks, William T. McLeod and Laurence Urdang. "Collins dictionary of the English language", *London: Collins*, c1986, 2nd ed., edited by Hanks, Patrick, **1** (1986).

[89]   S. Bandyopadhyay, M. Sengupta, S. Maiti, S. Dutta. "A survey of middleware for internet of things", *Recent Trends in Wireless and Mobile Networks*, pp. 288–296 (Springer, Berlin, 2011).

[90]   S. Bandyopadhyay, M. Sengupta, S. Maiti, S. Dutta. "Role of middleware for internet of things: a study", *International Journal of Computer Science & Engineering Survey (IJCSES)*, **2(3)**, pp. 94–105 (2011).

[91]   M. Chaqfeh, and N. Mohamed. "Challenges in middleware solutions for the internet of things", *Collaboration Technologies and Systems (CTS), 2012 International Conference on*, pp. 21–26 (IEEE, Piscataway, NJ, 2012).

[92]   L. Atzori, A. Iera, G. Morabito. "The internet of things: a survey", *Computer Networks*, **54(15)**, pp. 2787–2805 (2010).

[93]   T. Luckenbach, P. Gober, S. Arbanowski, A. Kotsopoulos, K. Kim. "TinyREST – a protocol for integrating sensor networks into the internet", *The proceedings are published as a Swedish Institute of Computer Science (SICS) technical report, SICS Technical Report* T2005:09.

[94]   University of Portsmouth Library. "Discovery service" [Online]. Available: http://www.port.ac.uk/library/infores/discovery/filetodownload, 170883,en.xls (2015) (Visited on 07/14/2015).

[95]   European FP7 ICT IP "Advanced Sensors and lightweight Programmable middleware for Innovative Rfid Enterprise applications" (ASPIRE) 2008–2010.

[96]   V.-M. Scuturici, S. Surdu, Y. Gripay, J.-M. Petit. "UBIWARE: web-based dynamic data and service management platform for AMI", *Proceedings of the Posters and Demo Track*, p. 11 (ACM, New York, NY, 2012).

[97]   A. Naumenko, A. Katasonov, V. Terziyan. "A security framework for smart ubiquitous industrial resources", *Enterprise Interoperability II*, pp. 183–194 (Springer, Berlin, 2007).

[98]   V. Terziyan, O. Kaykova, D. Zhovtobryukh. "UBIROAD: semantic middleware for context-aware smart road environments", *Internet and Web Applications and Services (ICIW), 2010 Fifth International Conference on*, pp. 295–302 (IEEE, Piscataway, NJ, 2010).

[99]   M. Caporuscio, P.-G. Raverdy, V. Issarny. "UBISOAP: a service-oriented middleware for ubiquitous networking", *Services Computing, IEEE Transactions on*, **5(1)**, pp. 86–98 (2012).

[100]   R. J. C. Benito, D. G. Márquez, P. P. Tron, R. R. Castro, N. S. Martín, J. L. S. Martín. "SMEPP: a secure middleware for embedded P2P", *Proceedings of ICT-MobileSummit*, **9** (2009).

[101]   L. M. S. De Souza, P. Spiess, D. Guinard, M. Köhler, S. Karnouskos, D. Savio. "SOCRADES: a web service based shop floor integration infrastructure", *The Internet of Things*, pp. 50–67 (Springer, Berlin, 2008).

[102]   Dogan Yazar and Adam Dunkels. "Efficient application integration in IP-based sensor networks", *Proceedings of the First ACM Workshop on Embedded Sensing Systems for Energy-Efficiency in Buildings*, pp. 43–48 (ACM, New York, NY, 2009).

[103]   H. Bohn, A. Bobek, F. Golatowski. "SIRENA – service infrastructure for real-time embedded networked devices: a service oriented framework for different domains", *Networking, International Conference on Systems and International Conference on Mobile Communications and Learning Technologies, 2006. ICN/ICONS/MCL 2006. International Conference on*, pp. 43–43 (IEEE, Piscataway, NJ, 2006).

[104]   D. Giusto, A. Iera, G. Morabito, L. Atzori. *The Internet of Things: 20th Tyrrhenian Workshop on Digital Communications* (Springer Science & Business Media, Berlin, 2010).

[105]   K. Aberer, M. Hauswirth and Ali Salehi. "Middleware support for the 'Internet of Things'", *In Proceedings of the 5th GI/ITG KuVS Fachgespräch* "Drahtlose Sensornetze" (2006).

[106]   C. Perera, P. P. Jayaraman, A. Zaslavsky, D. Georgakopoulos, P. Christen. "MOSDEN: an Internet of Things middleware for resource constrained mobile devices", *System Sciences (HICSS), 2014 47th Hawaii International Conference on*, pp. 1053–1062 (IEEE, Piscataway, NJ, 2014).

[107]   S. Hasan, E. Curry. "Thingsonomy: tackling variety in internet of things events", *Internet Computing, IEEE*, **19(2)**, pp. 10–18 (2015).

[108]   B. Billet, V. Issarny. "Dioptase: a distributed data streaming middleware for the future web of things", *Journal of Internet Services and Applications*, **5(1)**, pp. 1–19 (2014).

[109]   D. Conzon, T. Bolognesi, P. Brizzi, *et al.* "The VIRTUS middleware: an XMPP-based architecture for secure IoT communications", *Computer Communications and Networks (ICCCN), 2012 21st International Conference on*, pp. 1–6 (IEEE, Piscataway, NJ, 2012).

[110]   M. Eisenhauer, P. Rosengren, P. Antolin. "A development platform for integrating wireless devices and sensors into ambient intelligence systems", *Sensor, Mesh and Ad Hoc Communications and Networks Workshops, 2009. SECON Workshops '09. Sixth Annual IEEE Communications Society Conference on*, pp. 1–3 (IEEE, Piscataway, NJ, 2009).

[111]   A. O. Adetoye, A. Badii. "Foundations and applications of security analysis", *Chapter A Policy Model for Secure Information Flow*, pp. 1–17 (Springer-Verlag, Berlin, 2009) [Online]. Available: http://dx.doi.org/10.1007/978-3-642-03459-6_1. (Visited on 04/04/2016).

[112]   Linksmart.eu. "Linksmart middleware portal" [Online]. Available: https://linksmart.eu/redmine (2015) (Visited on 07/09/2015).

[113]   M. E. A. Gudgin. "SOAP version 1.2 part 1: messaging framework", *Recommendation, W3C* (June 2003) [Online]. Available: http://www.w3. org/TR/2003/REC-soap12-part1-20030624/. (Visited on 04/04/2016).

[114]   T. E. A. Bray. "Extensible markup language (XML) 1.0", *Recommendation, W3C* (February 2004) [Online]. Available: http://www.w3.org/TR/REC-xml. (Visited on 04/04/2016).

[115]   Linksmart. "eu.linksmart.security.communicationsecuritymanager. sym – linksmart open source middleware – linksmart middleware portal"

[Online]. Available: https://linksmart.eu/redmine/projects/linksmart-opensource/wiki/Eulinksmartsecuritycommunicationsecuritymanagersym (2015) (Visited on 06/09/ 2015).

[116] B. Dournaee, B. Dournee. *XML Security* (McGraw-Hill, New York, NY, 2002).

[117] M. Binna. "www.w3.org/2008/xmlsec/papers/c14n2_performance_ evaluation_thesis.pdf" [Online]. Available: http://www.w3.org/2008/xmlsec/ papers/C14N2_Performance_Evaluation_Thesis.pdf (2008) (Visited on 06/09/2015).

[118] L. Lan, B. Wang, L. Zhang, R. Shi, F. Li. "An event-driven service-oriented architecture for internet of things service execution", *International Journal of Online Engineering (iJOE)*, **11(2)**, pp. 4–8 (2015).

[119] T. Levendovszky, A. Dubey, W. R. Otte, *et al.* "Distributed real-time man-aged systems: a model-driven distributed secure information architecture platform for managed embedded systems", *Software, IEEE*, **31(2)**, pp. 62–69 (2014).

[120] G. A. Agha. "Actors: a model of concurrent computation in distributed systems", Technical report, DTIC Document (1985).

[121] A. Iivari, T. Väisänen, M. Ben Alaya, T. Riipinen, T. Monteil. "Harnessing XMPP for machine-to-machine communications and pervasive applica-tions.", *Journal of Communications Software & Systems*, **10(3)** (2014).

[122] Z. Ji, I. Ganchev, M. O'Droma, L. Zhao, X. Zhang. "A cloud-based car parking middleware for IoT-based smart cities: design and implementation", *Sensors*, **14(12)**, pp. 22372–22393 (2014).

[123] C. H. Liu, B. Yang, T. Liu. "Efficient naming, addressing and profile services in Internet-of-Things sensory environments", *Ad Hoc Networks*, **18**, pp. 85–101 (2014).

[124] ETSI. "ETSI – M2M" [Online]. Available: http://www.etsi.org/technologies-clusters/technologies/m2m (2015) (Visited on 07/08/2015).

[125] W. Zhiliang, Y. Yi, W. Lu, W. Wei. "A SOA-based IoT communication mid-dleware", *Mechatronic Science, Electric Engineering and Computer (MEC), 2011 International Conference on*, pp. 2555–2558 (IEEE, Piscataway, NJ, 2011).

[126] D. Conzon, T. Bolognesi, P. Brizzi, *et al.* "The VIRTUS middleware: an XMPP-based architecture for secure IoT communications", *Computer Com-munications and Networks (ICCCN), 2012 21st International Conference on*, pp. 1–6 (IEEE, Piscataway, NJ, 2012).

Chapter 2

Privacy in the Internet of Things

Santiago Suppan and Jorge Cuéllar

Summary

Privacy protection is gaining public attention, and even more so with the advances of the Internet of Things (IoT). But there is still no consensus on how privacy can be engineered, even for traditional Information Technology (IT) systems. This chapter, after discussing the main challenges, proposes a general methodology to manage the privacy engineering in the context of IoT and some technical building blocks and solutions.

More concretely, this chapter explains why IoT both increases the likelihood and aggravates the impact of data breaches in IT systems. Then, after presenting existing privacy protection principles and frameworks and mapping them to privacy protection mechanisms, a privacy development life cycle (PDLC) is proposed. Finally, privacy-enhancing technologies specific for the IoT are presented, as the application of a PDLC in IoT would require new and adapted technologies due to IoT's constraints.

2.1 Introduction

The Internet of Things (IoT) is expected to bring huge changes upon society and economy over the next 10–15 years. In doing so, it will also massively collect, store, and automatically process data, much of which will be related to individuals. In the Mauritius Declaration on the Internet of Things, see Reference 1, Data Protection and Privacy Commissioners underline that IoT can change the way we do things, e.g., in health care, transportation, and energy, but it can also reveal intimate details of users through the data collected by sensors. If this data is collected without the consent of the user, or leaks to attackers, significant consequences will follow for users and industries. The International Conference of Data Protection and Privacy Commissioners state therefore that providing the protection of big data in one system is a major problem already, the protection of data generated autonomously and ubiquitously from many IoT devices will be many times larger. More than 16 billion devices, see Reference 2, are forecasted to interact by 2020, allowing unexpected possibilities for users and industries. The potential of the IoT has been recognized worldwide, see, e.g., Reference 3.

In order to understand the privacy conflict in IoT, we have to understand how IoT can generate economical value. Kenneth Cukier, editor of *The Economist* and previous editor of *Wall Street Journal Asia* expresses it as follows [4]: "Well, think about it. You have more information. You can do things that you couldn't do before. [...] The general idea is that instead of instructing a computer what do, we are going to simply throw data at the problem and tell the computer to figure it out for itself. [...] And this idea of machine learning is going everywhere."

Data is an asset of great value, as seen in the Big and Smart Data boom. IoT is a key technology to reach new data sources. IoT devices can measure and quantify entities in a way that was not possible before, e.g., real-time electrocardiographic data from individual persons and data from public spaces with a multitude of entities. Data can be stored in the cloud and linked with other data from other sensors or other systems. This "enriched" data can reveal structures that correspond to purposes not identified in the first place, which fundamentally contradict the data protection principle of consented and purpose-bound processing.

A common understanding in Big Data is that parties collecting the data (which are not the data subjects themselves) are free to decide what they want to do with it to achieve their business goals.

This again contradicts privacy principles: it is the data subject, that is, the person to whom the data refers to, and not simply the "data generator" or the "data controller", the entity that should determine how the data can be used and by whom.

The European Commission (EC) has expressed concern regarding the challenges of data protection in the IoT as well, see Reference 5. Particularly, the Article 29 Data Protection Working Party – an advisory working group of the EC – states in Reference 6 that the society's acceptance and the benefits promised to cities and industries regarding the potential of the IoT will heavily depend on the perception of data protection provided by IoT. The following quotation is taken from Reference 6:

Beyond legal and technical compliance, what is at stake is, in fact, the consequence it may have on society at large. Organisations which place privacy and data protection at the forefront of product development will be well placed to ensure that their goods and services respect the principles of privacy by design and are equipped with the privacy friendly defaults expected by EU citizens.

The remainder of this chapter is structured as follows: In Section 2.2, we give an overview of related literature. In Section 2.3, we look at why privacy is challenging and which consequences might happen if breaches occur. In Section 2.4, we discuss existing work on privacy principles and engineering for IoT. In Section 2.5, we describe how a privacy development life cycle (PDLC) could look like, how existing work can be integrated into the life cycle, and which phases still need further research. In Section 2.6, we present several building blocks and privacy-enhancing technologies (PETs) for IoT. We conclude the chapter with Section 2.7 with open challenges and future work.

2.2   Related work

This section gives an overview of work that is discussed throughout this chapter.

A wide variety of cryptographic and non-cryptographic PETs have been developed since the beginning of the world wide web. In References 7, 8, Goldberg *et al.* survey PETs for web systems over a time span of 10 years. One the one hand, the survey shows how once promising technologies have proven unuseful, e.g., re-mailers; on the other hand, it shows how proposals have matured into well-used privacy tools, e.g., onion routing.

Goldberg resumes his survey by saying that advances have been encouraging but much work remains. And even more so in IoT: most of the technologies surveyed, even mature ones, have to be adapted to match IoT's limitations. We discuss which limitations exist, see Section 2.6.7, and present several IoT adapted PETs in Section 2.6.

A development life cycle is supposed to help developers planning security from the design, iteratively build security mechanisms and risk reducing strategies as well as benchmark their systems. We relate in Section 2.5 to frameworks or methodologies of known Security Development Life cycles (SDLs) to achieve these properties, particularly Microsoft's SDL [9], the Building Security In Maturity Model, see Reference 10, and the Software Assurance Maturity Model (SAMM), see Reference 11.

For privacy there are several "manifestos" or "foundational principles", and a couple of books (like *Privacy Engineering* by Ian Oliver [12], *The Privacy Engineer's Manifesto* by Michelle Dennedy *et al.* [13]), which propose own methodologies based on the same or similar principles we discuss in Section 2.4. We see the need for an approach that is close to known SDLs with explicit suggestions on the methodology for each life cycle step instead of a broader questionnaire-based approach. We formulate our proposal in Section 2.5.

Foundational principles for privacy can be found in the privacy by design framework, see Reference 14, which has been suggested as the solution for data protection for several years now. We discuss its applicability for engineering privacy in Section 2.4.2.

Closely related to our proposal is the privacy-by-design methodology of PReparing Industry to Privacy-by-design by supporting its Application in REsearch (PRIPARE), see Reference 15. We discuss the PRIPARE method, similarities, and differences in Section 2.4.3.

We will also discuss other technologies in each section of the life cycle. Particularly, we propose to use the risk-based threat analysis approach of LINDDUN [16], privacy design strategies by Hoepman, see Reference 17, and the post-release emergencies play by the Canadian Office of the Privacy Commissioner (OPC), see Reference 18. We will discuss these methods in Section 2.5.

2.3   The challenges and consequences of privacy breaches

In Section 2.1, we pointed out a fundamental tension between IoT and privacy. In this section, we look at the possible consequences of data breaches and the challenges of applying data protection to future IoT systems.

Data is an asset of immense business value for companies as well as the trust that customers and users have in them. Privacy violations may endanger both, the value of the data and the trust that people have placed. Moreover, the costs of privacy-related lawsuits could be huge. Yet, privacy breaches and violations are common in the news. For instance in 2009, Facebook settled a privacy-related lawsuit over $9.5 million due to its service Beacon. Beacon tracked a user's behaviour at 44 external shopping websites and published the consumption activities of the user in its Facebook profile under the "News Feed", see Reference 19. Beacon was introduced in Facebook as an opt-out approach and needed no explicit user consent. Following the lawsuit, Facebook tried to implement further privacy controls and easier opt-out, see Reference 20, but nevertheless, Beacon was shut down shortly after that. Anthem Inc., second largest insurance company in the United States, leaked 80 million records including names, social security numbers, dates of birth, and other sensitive details such as health status [21]. Google Buzz is a further example. Buzz was a social online network that was supposed to compete with Facebook and Twitter [22]. Google took advantage of its Gmail user base and linked both services to generate a high user traffic. Based on the email traffic of the users, Buzz profiles showed user names and relationships between each other. The latter was the most intrusive property of Buzz, as email relationships of users were exposed publicly. Google was confronted with high criticism, including a lawsuit filed by The Federal Trade Commission, which was settled for $8.5 million, see Reference 23.

The consequences for companies after a privacy breach have been evaluated by several sources. For example, the Cost of Data Breach Study, see Reference 24, shows that a single lost or stolen record may cause an average damage of $154. Data breaches ranged in 2015 from 2.2 to more than 101 lost records per incident, equalling an economic damage from $338.8 to $15.554. This does not include "mega breaches" which, according to the study, disclose a minimum of 10 million records, causing a damage of 1.3 million US dollars or more. The costs per breach include direct and indirect costs. Direct costs incur engaging forensic experts, hotline support, and providing discounts to recompense the user. Indirect costs are extrapolated calculations of customer loss and diminished customer acquisition.

In some cases, the consequences for companies may turn out to be not as big as for the users. For example, the data breach of Sony's PlayStation Network in 2014 was estimated to cost more than $100 million, see Reference 25. In their following quarter financial report, see Reference 26, Sony stated the reduction of costs to $15 million. Additionally, Sony was able to use the attention from the breach to push their publicity, flagging the attack as an attempt to stop the release of a controversial movie produced by Sony, thus gaining more subscribers as Sony released the movie in their entertainment network exclusively, see Reference 27.

The Sony case shows that companies might not see the need for data protection, as they can keep the breach costs low. Therefore, to push companies to protect user data, high sanctions as well as regulations have been demanded that do not allow loopholes. These exist in current directives, as they are too high level to decide whether protection is adequate or not, see, e.g., the discussion in Reference 28. Accordingly, there is need for technology to protect personal data in different scenarios.

For example, if no efficient PET is available for IoT devices due to their constraints, only a best effort protection can be demanded from companies, even if they are insufficient.

The society can suffer from serious consequences besides economical loss if IoT is used to track users. One consequence can be the so-called *chilling effect*, see Reference 29. The chilling effect describes the fear of individuals of being judged by their actions and decisions. In order to avoid such effects and increase user acceptance, IoT system operators will have to be transparent to users, e.g., by showing that their systems are not used for extensive tracking. Privacy will therefore be a key acceptance criterion for IoT.

Technical means to support privacy in IoT could therefore be beneficial for users and service providers, but there are "huge challenges faced by Internet of Things developers, data protection authorities and individuals", see Reference 1.

For example, the data gathered in public spaces relates to many different individual data subjects, and it will not be easy to ask them for consent or inform them about the purpose. The transparency of the data provenance and integrity is difficult to guarantee in a scenario where subjects are continuously being monitored and tracked by a large number of devices. And even if technical solutions exist, constrained devices used in IoT make the use of security and privacy mechanisms difficult to implement, configure, and use.

In Section 2.6, we will look into details of which PETs are needed in IoT.

2.4 Privacy principles and engineering for the IoT

There are several well-established security engineering principles, best practices, and guidelines for software and system development. Frameworks or methodologies like the SDL, see Reference 9, the Building Security In Maturity Model, see Reference 10, and the SAMM, see Reference 11, can help developers to plan security from the design, iteratively build security mechanisms and risk reducing strategies as well as benchmark their systems.

In the case of privacy, the situation is not clear-cut. There are several key approaches, and several "manifestos" or "foundational principles", and a couple of books (like *Privacy Engineering* by Ian Oliver [12], *The Privacy Engineer's Manifesto* by Michelle Dennedy *et al.* [13]), but there is still no wide consensus on how a possible privacy development framework (PDLC) should look like, nor what the main engineering principles would be, see Reference 30. Foundational principles for privacy can be found in the privacy by design framework, see Reference 14, which we will briefly present later. Privacy engineering proposals can be found in the National Institute of Standards and Technology Standard 806 [31] and PRIPARE [15], and privacy guidelines in Spiekermann and Cranor [32]. According to the Internet Privacy Engineering Network [33], "one reason for the lack of attention to privacy issues in development is the lack of appropriate tools and best practices [...]. There are, unfortunately, few building blocks for privacy friendly applications and services [...]". We will present some building blocks and PETs for IoT in Section 2.5.3.

In this section, we present the current approaches to Privacy Engineering, including Privacy by Design and PRIPARE, and sketch how a PDLC should look like. But first we look at the Privacy Principles from a regulation point of view.

2.4.1   The European Data Protection Rules

The European Union (EU) directive of 1995 adopted many principles of fair information practice which can be found also in the Canadian Privacy by Design model. The principles can be found in two directives, Directive 95/46/EC and Directive 2002/58/EC. The directive of 2002 is an extension of the privacy protection rules of 1995 for privacy in electronic communications. In summary, the principles are:

Consent states that personal data shall be collected and processed only for a specified, explicit, and legitimate purpose. The words "specified" and "legitimate" imply that a data subject and the processing party have to agree on a common consent to how the data is exactly processed and which processes are outside of a legitimacy.

Purpose legitimacy and specification is closely related to Consent. This concept explicitly asks developers to understand which personal data is exactly being used how and for which purpose.

Collection limitation means that the collection of personal data must be "adequate, relevant and not excessive".

Data minimization is familiar with the previous principle. Data minimization heavily supports privacy by design, by helping to avoid the collection, generation, and storage of personal data.

Notice and Access is defined as "communication to him [i.e., the data subject] in an intelligible form of the data undergoing processing and of any available information as to their source, [and the] knowledge of the logic involved in any automatic processing of data concerning him at least in the case of the automated decisions [. . .]".

Individual Participation According to the European Directive 95/46/EC on Protection of Personal Data (art.12 (b)), data subjects have the right to withdraw from the processing of their personal data. This includes data collection.

Accountability Data subjects have the right to receive compensation from a processing party in case of data breaches. Therefore, data controllers have to be identified clearly and assigned responsibilities for the data they are controlling.

The principles express fundamental rights of all EU citizens and have helped to harmonize data protection regulation in Europe as they are widely applicable due to their neutrality in terms of technology. But, on the other hand, some aspects remain unspecific. For example, it is unclear how much choice and control citizens should have.

In Reference 34, Koops discusses the directives from a citizen perspective and observes that exercising data subject rights is

highly theoretical. Yes, you can be informed, if you know where to look [. . .].
Yes, you can request controllers [. . .] if you know that you have such a right
in the first place [. . .]. Yes, you can request correction or erasure, if you
know whom to ask (but how are you ever going to reach everyone in the
chain [. . .]?). There are simply too many ifs and buts to make data subject
rights meaningful in practice.

The new European Data Protection Regulations want to target the difficulty to apply the protection rules and the unclear scope by defining specific and additional regulations [35].

2.4.2 Privacy by Design

Privacy by Design, see Reference 36, has been suggested as the solution for data protection for several years now. Privacy by Design seeks to unify privacy methodologies, define processes via a "Privacy Impact Assessment Guidelines" and identify fundamental principles in a holistic framework. The core of the framework comprises seven fundamental principles which stress the importance of considering privacy from the early steps of design and through the whole life cycle. The steps are:

Be proactive not reactive Privacy should be included in a system preventively, that means in the design and architecture, not remedial.

Privacy as the default It has been shown that the default settings of a system are mostly used in the lifetime of a system, even if options are available. Therefore, the most privacy-preserving state of a system should be its default.

Privacy embedded into design This point re-enforces that privacy should be embedded into the design.

Full functionality Integrating privacy and security does not have to reduce the functionality or the utility of the system. There are many technologies to choose from for this purpose.

Full life cycle protection Data protection should be present for the whole life cycle of data.

Visibility and transparency Visibility and transparency have to be employed to create trust to users. Privacy by Design mentions three concepts: accountability (the responsibility for personal data should be clear and documented), openness (show what data you have and how you are processing them), and compliance (necessary steps to monitor, evaluate, and verify compliance with privacy policies and procedures should be present in a system).

Respect for user privacy User-centric development has been shown to increase the success of products; therefore, privacy protection should be regarded as a tool to make a system user-centric.

The Canadian model is considered foundational, but it needs additional frameworks to support system engineers due to its high-levelness, see, e.g., the opinion of Reference 37.

2.4.3   PRIPARE

PRIPARE, see Reference 15, aims to provide a privacy-by-design methodology and process reference model for systematically incorporating Privacy-by-Design in software engineering. PRIPARE has identified a lack of privacy practice that can be used through system engineering life cycle. Therefore, PRIPARE defines a methodology that includes processes, practices, and methodologies in order to integrate them to system engineering phases (analysis, design, implementation, verification, release, maintenance, and decommission).

The PRIPARE approach is goal oriented and a risk based. During the PRIPARE process, privacy requirements and goals are defined and integrated to other system goals such as those for functionality and security.

The PRIPARE methodology addresses a system's architecture from a privacy point of view as well as privacy requirements may affect the architecture changes.

PRIPARE structures its methodology in phases, which are (see Reference 15):

Analysis In this phase, the functional description of the system is specified and high-level privacy requirements are elicited. The goal of this phase is to understand what privacy controls must be implemented to effectively operationalize privacy in the system.

Design The design phase focuses on how the privacy controls have to be build. The definition should contain hardware and software architecture, components, modules, interfaces, and data flows.

Implementation In this phase, the system described in the design phase is implemented, following an architectural model and privacy-enhancing design principles.

Verification This phase ensures that the system meets privacy operational requirements. PRIPARE proposes to check implementation properties with formal verification, code reviews, and dynamic flow analysis. Furthermore, "posteriori" compliance controls are implemented to support accountability.

Release This phase defines several processes that have to be completed before system release. The processes include elaboration of an action plan to respond to the discovery of privacy breaches, creation of a system decommission plan, and a final privacy review.

Maintenance In this phase, a data controller has to react to privacy incidents and try to minimize the damage for affected subjects as much as possible. This phase requires immediate actions and a well-defined communication plan with subjects and authorities.

Decommission The purpose of this phase is to correctly dismantle the system according to applicable legislation and policies. The decommissioning of the system should not result in possibilities for data breaches.

PRIPARE's "privacy and security-by-design methodology" is not a life cycle itself, the methodology is structured in seven different phases in order to match those of common system engineering phases, see Reference 38. In Section 2.5, we propose a PDLC which could be merged with the PRIPARE methodology. Additionally, we present building blocks in Section 2.6 to give concrete technologies for the PDLC.

2.5   A privacy development life cycle

The term "System (or Software) Development Life cycle" describes a process for planning, creating, testing, and deploying an information system. More specifically, a security life cycle development was introduced as a systematic approach for security in software engineering. A PDLC should have similar goals, e.g., systematically introducing a privacy methodology in system engineering. Security and PDLCs will have significant differences, but if done in an integrated way, beneficial synergies form both approaches can be obtained.

The Microsoft SDL [39] and the Open Web Application Security Project (OWASP) SAMM [11] are two of the most popular SDLs primarily for software development. Overall, the seven steps[1] of both security life cycle frameworks are as follows:

1. Train personnel or ensure that personnel are qualified.
2. Identify threats, evaluate risks (which threat is going to be mitigated, which threats and risks will be simply accepted?), and elicit requirements.
3. Design the system according to the requirements.
4. Implement the system, fulfilling all requirements.
5. Verify if the system fulfils the requirements.
6. Deploy the system while making sure the requirements will still apply in the deployment environment.
7. Keep the system developers ready to respond to any conflicting or emerging situation.

In the following sections, we will propose a PDLC closely resembling the seven phases of the security life cycle framework. The phases differ slightly from their Software Development Life Cycle counterparts in name and content as they take into account privacy principles and other privacy-related work.

2.5.1   Education of system developers

As with security, training developers in privacy topics are necessary. Although all team members should understand why privacy protection is fundamental and be familiar with the main guiding rules (say, the EU Data Protection Rules or the applicable guidelines), we assume that at least one person in the team is particularly well trained in the technical aspects of designing and implementing privacy-friendly systems. We call him the "privacy expert" of the team. He should know a privacy engineering framework like PRIPARE. The most critical condition for achieving a privacy-friendly

---

[1]The initial steps *Strategy & Metrics and Policy & Compliance* of SAMM are not presented for the sake of simplicity.

product is the presence of one or several privacy experts in the team. The expert is responsible for data protection expertise in the development team and should be integrated in every phase of the life cycle. He also brings the knowledge where to find mature technical privacy solutions (PETs) and best practices. The life cycle itself does not focus on developing new technologies, which could cost a considerable amount of time, research, and technical expertise, but on using existing building blocks and suitable PETs. We give a brief overview of PETs for the IoT in Section 2.6. Particularly, the IoT requires specialized technology for computational and battery constraints. A privacy expert needs therefore continuous refinement of technical skills and state-of-the-art knowledge.

Legal support will probably be of need to resolve privacy-related emergencies. In these situations, a privacy expert should be aware when legal support is required.

2.5.2   *Phase 1 – purpose definition and data minimization*

The first phase of the privacy life cycle development is the specification of requirements for the system. Here, the system's functional requirements are analysed by posing the following questions, which follow the principle of *Purpose*: *Specifically, what personal data does the system collect? What is their specific purpose? Can the system reach its desired functionality with less personal data?* The following process is iterated to stepwise obtain more concrete and operational privacy requirements or PETs.

1. Obtain or define the system data flow and the system's functional goals and requirements.
2. Determine which personal data is needed to achieve the system's functional goals.
3. Analyse the functional requirements and determine if existing PETs can help to minimize the data that is needed for the system.
4. Determine the limits of data usage and data retention in the system (say, data is deleted after 2 weeks).
5. The privacy expert analyses the proposed solution and suggests new possible technologies to reduce data usage in the system.

2.5.3   *Phase 2 – threats and risks evaluation*

After the definition of the required personal data in the system, privacy requirements, and privacy goals, this phase is used for privacy threat analysis. Several frameworks for privacy threat analysis have been proposed, such as LINDDUN [16], PriS [40], and FPFSD [32].

LINDDUN is especially well suited for the integration of a PDLC as it is based on STRIDE, see Reference 41, part of the SDL. System developers trained in SDL should be able to learn the LINDDUN method easily, reuse existing system models (particularly data flow diagrams or "DFDs") for their systems and see synergies or problems of both security and privacy goals. LINDDUN follows STRIDE in defining six steps. The first three cover the "problem space", focusing on the problems, identifying privacy threats, and defining requirements of the system. The last three steps cover the "solution space" which aim at fulfilling the requirements, see Reference 16:

Define DFDs of the system In this step, a graphical representation of the information flow in the system is created. This step is equal to the step in the STRIDE methodology and could be combined with LINDDUN.

Map privacy threats to DFD elements In this step, system components are mapped to privacy threats. LINDDUN defines seven threat categories[2]: *Linkability* is the property of linking two or more actions/identities/pieces of information, *Identifiability* is the property of linking the identity and an action or information, *Non-repudiation* is the inability to deny a claim, *Detectability* is the property of being able to distinguish sufficiently whether an entity exists or not, *Information Disclosure* is the property of revealing confidential information, *Unawareness* is the property of not knowing the consequences of sharing information, and *Non-compliance* is the property of not being compliant with legislation, regulations, and policies.

Identify threat scenarios LINDDUN provides so-called threat trees to identify threat scenarios. The privacy analyst should examine each of the branches of the tree with a specified DFD element in mind.

Prioritize threats/risk analysis All the potential privacy threats that are suggested by the privacy threat trees are evaluated and prioritized via risk assessment.

Elicit mitigation strategies The suitable mitigation strategy for each thread is determined.

Select PETs The classification of PETs according to the mitigation strategies to which they adhere enables a more focused selection of suitable privacy-enhancing solutions.

LINDDUN also supports the integration of any risk assessment framework, for instance the one the security team might in the SDL.

2.5.4 *Phase 3 – design*

The design phase develops strategies for implementation, verification, release, and response. In this phase also, functional, security, and privacy requirements are adjusted to one another. For example, functional requirements might need to change to respect policies, security procedures might need to be adapted to support unlinkability, and privacy requirements might turn out impractical due to core functional requirements and need to be reshaped.

Conflicts might appear between goals, therefore best practices can be useful. Best practices are strategies that have been employed by others with good results.

---

[2]The acronym LINDDUN stems from the seven categories.

For example, Hoepman, see Reference 17, has defined eight design strategies for privacy which can be realized using privacy patterns (i.e., best practice solutions), namely:

Minimize states that the amount of personal data that is processed should be restricted to the minimal amount possible. This is the most basic privacy design strategy.

Hide states that any personal data, and their interrelationships, should be hidden from plain view.

Separate states that personal data should be processed in a distributed fashion, in separate compartments whenever possible.

Aggregate states that personal data should be processed at the highest level of aggregation and with the least possible detail in which it is (still) useful.[3]

Inform corresponds to the important notion of transparency. Data subjects should be adequately informed whenever personal data is processed.

Control states that data subjects should be provided agency over the processing of their personal data.

Enforce A privacy policy, compatible with legal requirements, should be in place and should be enforced.

Demonstrate requires a data controller to be able to demonstrate compliance with the privacy policy and any applicable legal requirements.

In Reference 17, Hoepman provides a set of patterns to each strategy to support their technical engineering. As some strategies have rarely been used before, Hoepman points out that new patterns are needed. The reader is referred to the privacy patterns database [43].

At this point, we would like to add the following principles to the ones just mentioned:

Early application of policies and filtering The processing of personal data should be in the devices under the control of the data subject or if that is not possible, at the earliest point of its generation. This strategy takes advantage of increased processing power in personal devices. Spiekermann and Cranor, see Reference 32, describe how this strategy can eliminate the need for data transfer and remote storage, minimizing unwanted secondary data use, compliance with policies, and compliance with consented agreements.

Do not link Hoepman describes a separation strategy to process data in a distributed fashion, but e.g., storing data in separated databases is not enough, if it can be re-linked across the databases. This strategy helps to avoid such cases establishing a mechanism that actively checks for possible identifiers and allows proper separation without the possibility to re-aggregate the data.

Usability and data-flow transparency Ensuring the usability in privacy controls has several objectives: make privacy controls usable for a variety of users with

---

[3]The reader should be warned that the word aggregation is used in the privacy context with different meanings: for instance Solove, see Reference 42, uses the word to describe "gathering and combination of various pieces of data about a person" and in that sense aggregation is a privacy threat. Here, it means abstraction or replacement by more general statistics on the data.

different skill levels, integrate privacy controls seamlessly into the system, and make the users understand what they are seeing and what they can invoke with the controls provided to them.

Rubinstein and Good propose in Reference 30 to use field studies, interviews, surveys, and related methods to understand the user requirements, pain points, and expectations, for the creation of narratives that help to drive software engineering requirements, which are then incorporated into the overall development plan. The narratives or scenarios should be transparent to the user, allowing him to visualize how his personal data is used and how it flows in the system.

2.5.5 *Phase 4 – implementation*

Proper documentation and by-default configuration are key in this phase. Users must be able to perform informed decisions in a privacy-preserving way without much trouble. In other words, the system should behave privacy-friendly out of the box. This is called "privacy-by-default" and is one of the most important fair information and privacy by design principles, as the majority of users will interact with a system in its lifetime with the default settings, as pointed out in Reference 44.

"Secure Coding" procedures will be needed to avoid privacy issues, which could otherwise become visible later in the system. PETs need to be securely implemented in the same way as security mechanisms, e.g., by coding experts and verified with code reviews. Implementation strategies, as defined in phase 3, help to assure that the implementation effort is controllable, timely and reaches the desired quality. Software developers and privacy experts should closely work in this phase to avoid problems such as an improper choice of libraries with unwanted effects (like the use of logging of data including personal information, or the presence of vulnerabilities or leaks).

2.5.6 *Phase 5 – verification*

It remains unclear if testing procedures, such as pen-testing, fuzzing, etc., can be used for privacy purposes.

But the methods used in code review offer also good information about data and information flow in programs and about the presence and enforcement of privacy-enhancing mechanisms.

2.5.7 *Phase 6 – release of system and education of stakeholders*

Phase 6 is used to develop strategies in case that vulnerabilities are discovered on release. These strategies are carried over to the next phase [39]. Strategies cover assignment of responsibilities, emergency response methods and emergency assessments, technical actions, and communication strategies. Privacy cannot be protected simply by technical components and this holds for security as well. The education of system stakeholders takes a significant role in this phase. Stakeholders of the system are system administrators and operators and system end users. Operational stakeholders need to know which data is processed by the system and what kind of implications this might have for users. Technical protection might be useless in certain scenarios that might seem unlikely, yet the operators should know them to be able to react in case they occur.

Data subjects need to be informed about how their data is processed and which tools they are provided to exercise their privacy rights. The released system should be accompanied by an according privacy disclosure which describes the system's use of personal data, documentation which details the tools that the system provide and user communication tools like a "quick text" to address likely user questions.

Users, and in particular also system administrators and other personnel that may interact with the system, need to be informed about how their actions affect the privacy of others and which actions can lead to privacy violations.

SDL proposes to validate the system's privacy standards by a privacy advisor or a privacy seal of quality prior to release. A legal privacy expert should review the documents and overview the release process.

2.5.8   Phase 7 – response

The last phase is one of the most significant in the life cycle. It carries over the results from the release phase for rapid response strategies.

Breaches might have a significant impact, as discussed in Section 2.3, and the team must be prepared to respond efficiently and timely to them, as they can occur unexpectedly. A response team must therefore develop a response plan that includes preparations for potential post-release emergencies. The OPC proposes, in Reference 18, four steps for this phase:

Breach containment and preliminary assessment In this step, immediate actions to stop the breach are carried out, the investigation leader and a response team are assigned. Legal action against the attackers is suggested as well.

Evaluation of the risks associated with the breach In this step, the risks associated with the breach are evaluated and first actions are triggered. The risk depends on the amount, sensitivity, and context of the compromised data, e.g., if the data was encrypted or not and if identifiers or other information links them to particulars. Assessments can help to identify the individuals affected, the root cause and the foreseen harm and find adequate mitigation strategies.

Notification In this step, the users are notified of the possible consequences the breach might have. The notification should be as soon as (reasonably) possible and personal, by phone, email, etc. In this step, also further organizations can be informed, such as cyber defence centres, credit card companies (if credit card data was stolen), etc.

Prevention In this step, a prevention plan is defined. The OPC suggests the level of effort should reflect whether it was a systemic breach or an isolated instance.

These steps aim at fast communication and support strategies between companies and users. They help the users to understand what possible consequences the breach may have and give them a transparent view of the emergency response strategies form the company.

A legal support might be needed to handle consequences, but also to initiate legal actions against the attackers. A root-cause mitigation team investigates why a breach was possible and develops a mitigation plan that has to be realized rapidly.

2.6   PETs for the IoT

Considerable expertise is needed to apply a PDLC and to identify the right PETs for a system. The IoT brings additional challenges due to its constrained and heterogeneous environment which would require an additional adoption of existing PETs to IoT.

In this section, we give a short overview of the findings of the project REliable, Resilient and secUre IoT for sMart city applications (RERUM) in respect to suitable PETs for the IoT. The PETs support different privacy principles and help to mitigate several threats (we refer to the principles of the current European Data Protection Rules as seen in Section 2.4.1, and the threats discussed in Section 2.6).

2.6.1   Privacy Dashboard

Dashboard is used as business intelligence tool to represent and analyse data. A Privacy Dashboard is a graphical interface which enables the management of policies for the private information of users, see References 46, 47. The dashboard supports the principles of Individual Participation, Notice and Access and allows a user to know where his data is, so he may delete and update his data or stop the data collection at any time. Since it is impossible to expect that all users of an IoT system have strong technical background, it is not viable for them to express their privacy policies in a complex policy language. The graphical interface visualizes how a user's devices behave in the system and it allows changing that behaviour according to the user's preferences. These preferences are converted (automatically) to policies in a machine-readable format. Additionally, the Privacy Dashboard helps to understand what kind of data their devices are sharing with the IoT system, thus avoiding that users are overwhelmed with raw data, i.e., without knowing what the data really means.

2.6.2   Consent management

A consent manager is a dedicated component of an IoT system that allows a data subject to see which applications are requesting access to his personal data and the purpose of the request. A given or denied consent is stored in a consent database, thus retrievable for accountability issues. A data subject may declare his consent to a service provider or withdraw a previous one at any time, thus supporting Consent, Notice and Access, Participation, and Accountability.

The consent manager is related to an authorization engine, where the final decision is made by the subject itself. A full description can be found in References 47, 48.

2.6.3   Support of user defined policies

In order to support the privacy principles of Purpose Legitimacy and Specification, and Individual Participation, the user should have a way of expressing his privacy choices to the system, in the form of user-defined privacy policies. It is of paramount importance that the policies are easy to retrieve when needed. This is not so easy to guarantee in an IoT-based system with huge amounts of data. For this purpose, the policies should be linked to physical entities they relate to, when the data is at rest. Data in transit should travel with so-called sticky policies.

2.6.3.1   Privacy policies

For the following PETs, we refer to the IoT Architecture Reference Model (abbreviated "IoT-ARM"), see Reference 49, to elaborate how physical objects are represented in IoT. The IoT-ARM refers to physical objects as Physical Entities. Physical Entities can be any object in the "real world" including living things. Physical Entities are represented as Virtual Entities in the virtual space. The relationship between Physical and Virtual Entities is defined as a conceptual entity called Augmented Entity. If an Augmented Entity exists, the corresponding Virtual Entity changes according to the Physical Entity. If the Physical Entity changes physically, the values of the Virtual Entity are updated. The Virtual Entity is only be updateable if the Physical Entity can change accordingly. Virtual Entities expose resources that can be requested by clients, e.g., service providers.

A natural way of binding the privacy policies to a Physical Entity is to link them to the entity in its virtual representation. The policies can now be enforced on access requests for a Physical Entity, as the requests have to be directed to the virtual representation. A full description of the integration of privacy polices within the IoT domain model can be found in Reference 48.

2.6.3.2   Sticky policies

Sticky policies are privacy policies that are attached to data and accompany it whether it is stored or in transit. These policies promote the user's wishes of allowed actions and consent obligations for parties processing the data. In the IoT-ARM, sticky polices can be located on the Virtual Entities as well. The reason is that the Virtual Entities are the earliest point of data creation and that policies can be stuck to the data as soon as it is created. The interested reader is referred to References 50–52 for technical details and for a full description of the integration of sticky polices within the IoT domain model, please refer to References 48, 53.

2.6.3.3   Temporary suspension of consent

A user should be able to suspend his consent any time and therefore to intervene and deactivate the collection of data from any of his devices any time. Such a mechanism is deeply rooted in an IoT system as it has to deactive the data collection of devices independently of the way they are sending data. The integration of such a component in the IoT-ARM needs further elaboration of the architecture. We refer to the "Activator and Deactivator of Data Collection" description in Reference 48 for details.

2.6.4 *Privacy-enhanced authentication and authorization*

Privacy-enhanced authentication in IoT refers to a set of mechanisms that allow integrity and authentication of users, devices, and gathered data. An example of a privacy-enhancing authentication mechanism is the Group Signatures scheme, see Reference 54. The scheme authenticates the group of potential signature creators instead of the exact entity signing the data. Group signatures are an identity-based authentication scheme that help the principle of Data Minimization.

A more modern privacy-enhanced authentication scheme is Attribute-Based Credentials, described in ABC4Trust, see Reference 55. The ABC4Trust architecture defines a common mechanism to query only the minimal information required by a service provider.

2.6.5 *Pseudonym management systems*

A pseudonym system supports Data Minimization by hiding the identities of devices and users from the services and other system participants, if they are not necessarily needed. In cases where attackers or intruders are able to steal records from databases, pseudonyms will prevent that individuals are tracked down through their identities.

Pseudonym systems can be categorized in three concepts, see Reference 56. Spatial concepts are based on mix-zones, where pseudonyms are exchanged when the pseudonymized participants meet. Time-related concepts change pseudonyms after a certain time. User-oriented concepts allow the users to decide when their identity should be changed. The decision can be based on the user's own policies and thresholds for the automatic pseudonym change.

2.6.6 *Location privacy*

Geo-location PETs support Data Minimization in traffic applications. They enable the system to send the minimal amount of information to location-based service providers. The methods vary according to their scenarios: pseudonym exchange technologies are used in vehicular ad hoc networks, see Reference 57. Data-based technologies are used in floating car observation. Data obfuscation and regular pseudonym exchange can be combined to create a strong obfuscation of the association of users and location. This is very important, as the tracking of location information discloses a large amount of information about the habits, activities, and preferences of subjects.

2.6.7 *PETs for constrained environments in IoT*

The Internet Engineering Task Force (IETF) has classified devices by their computation and storage capacities in Reference 58. The most used devices in IoT are "class 1" with a working memory restriction of *10 KB RAM* and a storage capacity of *100 KB Flash*. Class 1 devices are typically powered by coin or dry cell batteries with a maximum capacity of *2 376 joules*, see Reference 59.

The classes are not clear cut and other classes are used in some cases for IoT. For the elicitation of adequate PETs, the constraints of class 1 are those that will impose a significant factor. We will look into details in Section 2.5.3.

Smart objects in the IoT range from desktop computers to small hardware embedded in everyday objects such as clothing. These small computers are constrained in computing power and memory and therefore make it difficult to integrate state-of-the-art privacy technologies. In the following sections, we look at two common requirements, namely authorization and authentication, and exemplify two currently researched approaches that allow those requirements to be fulfilled.

2.6.7.1   Efficient privacy-friendly authorization

Many services will consume data from IoT devices and, accordingly, the authorization of those services against the devices will be needed. Authorization frameworks such as Security Assertion Markup Language (SAML) and OAUTH are based on tickets/tokens, which require an authorization party to fully understand the token and deny or allow the request based on policies. A device with 10 KB RAM and a storage capacity of 100 KB would be very constrained in managing the tokens, policies and its functional processes for measurement, actuation, etc. IoT devices will need therefore a light-way approach for authentication.

The basic idea is to *delegate* the "heavy" computational effort to another component, that means, efforts for policy validation, access decision, and token generation. An access request from a service provider to a device is delegated (by the device) to its "authorization manager". The authorization manager is an entity that knows the device's resources and policies, and decides upon access or denial of the request. If the decision is positive, a token/ticket is generated and sent back to the service provider. The token/ticket information is generated in such a way that the device can unequivocally identify an access permission for a respective request. The service provider sends a new request with the token/ticket to the device. If the request matches the ticket/token information, the request is granted.

This approach is depicted in Delegated CoAP Authentication and Authorization Framework (DCAF) [60]. The DCAF architecture defines a constrained device as "server", as the device senses data and "serves" it to service providers and clients. Accordingly a "server authorization manager" (SAM) is defined, which is the delegation component assigned to the server.

A client or service provider has to obtain a ticket or token from the SAM[4] first to request data from the server. The server has a whitelist approach. If the ticket or token is known to the server and the request matches the token, the access is granted, else, the access is denied.

However, building token material that does not identify a client every time it wants to consume data from a server is further challenge. As delegation takes place, PETs have to be integrated in both components, server and SAM. An extension with privacy-friendly token material is therefore proposed in Reference 61. Here, the authentication between SAM and client is done in a privacy-enhanced way, e.g., group signatures, and the tokens contain constantly changing secrets. The secrets are

---

[4]Ticket or token depends on the underlying authorization protocol.

generated in such a way that the constrained server is still able to identify them by itself (even if they are often changing), with only little resources.

Both approaches are actively being developed in the IETF Authentication and Authorization for Constrained Environments (ACE) working group. They show how delegation can help to overcome computational constraints and how it might be a building technology for further PETs.

2.6.8   Trust in IoT

In Section 2.6.4, access control was presented as an important PETs for IoT. But compliance with access control policies cannot be forced and relies often on the expected behaviour of the parties involved. Trust and reputation are highly related to security, but they can also contribute to other factors such as reliability, availability, and privacy, see Reference 62. In this section, we look at the role of Trust and Privacy in IoT and particularly why access control needs to be based on trust to be adequate for Privacy in IoT.

Gambetta defines Trust as the subjective probability by which an individual, A, expects that another individual, B, performs a given action on which its welfare depends, see Reference 63. In web systems, trust is often a binary probability that is linked to authentication. In a typical access control scenario, a user is fully trusted if he is able to authenticate against a service. The service is then able to decide on the user's requests, because the service knows the role he is assigned to. Newer access control methods, such as single-sign-on, do a variation of authentication, and role assignment, but the underlying method is the same: a user is trusted if he is able to authenticate and his role is known to the system.

In IoT, users are not predetermined and thus cannot be known a priori by the services. Services themselves might be created dynamically including their own access control components. Additionally, services might be resource constrained and thus not able to store a high amount of user IDs and user roles.

The assignment of rights to users has to be done dynamically, that means, if an unknown user wants to access a system, a mechanism has to reason about if the user is allowed to access a service or not. This can be done through delegation and trust management, as proposed in Reference 64. A user that is trusted by the system can delegate all or a subset of his rights to another user that he trusts and that can fulfil the requirements associated to the rights. Delegated rights can be access- and delegation rights. With delegation rights, the users can generate a trust chain. Every member of the chain can access the system according to his respective rights until they expire or until the chain is broken. This is the case when a user becomes untrusted or fails to meet the requirements associated with a delegated right. In that case, every user in the chain will not be able to exercise his rights, and the chain has to be created anew.

How users and the service quantify trust may be different? Depending on the context, they might share secrets that define their trustworthiness or they might evaluate trust based on reputation. Reputation is the collected and processed information about a user's former behaviour as experienced by others. This is especially interesting due to the heterogeneity of users in IoT.

2.7 Discussion

In this chapter, we presented the challenges of privacy in IoT, gave an overview of the privacy engineering methodology, and proposed a privacy life cycle development. We referred to the overall lack of PETs and building blocks in privacy in IoT, and presented some building blocks and their integration in an IoT architecture. Additionally, we introduced technologies for efficient PETs that are currently being researched and pointed out that these technologies can help to build further PETs for constrained devices.

Engineering privacy in IoT remains difficult because privacy engineering itself is a task that still needs to be well understood. Privacy engineering lacks a consensus of best practices and supporting tools. We see the need for more efficient PETs suited for constrained environments and in particular tools that support transparency to users allowing them to control and access their data throughout multiple providers and processes.

Acknowledgements

The research leading to some of the results presented in this chapter has received funding from the European Union's Seventh Programme for research, technological development, and demonstration under grant agreement no. 609094. We would also like to thank the RERUM project partners for the fruitful discussions.

References

[1] Kohnstamm J, Madhub D. "Mauritius Declaration on the Internet of Things". In: *36th International Conference of Data Protection and Privacy Commissioners*; 2014. Available from: http://www.privacyconference2014. org/media/16596/Mauritius-Declaration.pdf. (Visited on 04/04/2016).

[2] Vermesan O, Friess P, Guillemin P, *et al.* "Internet of Things strategic research roadmap". *European Research Cluster on the Internet of Things*; 2011. Available from: http://internet-of-things-research.eu/pdf/IoT_Cluster_ Strategic_Research_Agenda_2011.pdf. (Visited on 04/04/2016).

[3] Sundmaeker H, Guillemin P, Friess P, Woelfflé S. "Vision and challenges for realising the Internet of Things". *EUR-OP*; 2010. Available from: http:// www.internet-of-things-research.eu/pdf/IoT_Clusterbook_March_ 2010.pdf. (Visited on 04/04/2016).

[4] Cukier K. "Big data is better data". *Electronic (TED)*; 2014. Available from: https://www.ted.com/talks/kenneth_cukier_big_data_is_better_data? nolanguage=e. (Visited on 04/04/2016).

[5] Internet of Things Expert Group (IoT-EG). "IoT privacy, data protection, information security. *Electronic*; 2013. Available from: http://ec.europa.eu/ information_society/newsroom/cf/dae/document.cfm?doc_id=1753. (Visited on 04/04/2016).

[6] Party ADPW. "Opinion 8/2014 on the recent developments on the Internet of Things". 14/EN WP 223; 2014.

[7] Goldberg I, Wagner D, Brewer E. "Privacy-enhancing technologies for the Internet". *DTIC Document*; 1997.

[8] Goldberg I. "Privacy-enhancing technologies for the Internet, II: five years later". In: *Privacy Enhancing Technologies*. Berlin: Springer; 2003. p. 1–12.

[9] Howard M, Lipner S. *The Security Development Lifecycle*. Sebastopol, CA: O'Reilly Media, Incorporated; 2009.

[10] McGraw G, Chess B, Migues S. *Building Security In Maturity Model*. San Francisco, CA: Creative Commons; 2009.

[11] Chandra P, Deleersnyder S. *OWASP Software Assurance Maturity Model*. Mountain View, CA: Creative Commons (CC) Attribution-Share Alike; 2012.

[12] Oliver DI. *Privacy Engineering: A Dataflow and Ontological Approach*, 1st ed. North Charleston, SC: CreateSpace Independent Publishing Platform; 2014.

[13] Dennedy MF, Fox J, Finneran TR. *The Privacy Engineer's Manifesto: Getting from Policy to Code to QA to Value*, 1st ed. New York, NY: Apress; 2014.

[14] Cavoukian A. "Privacy by design". *Electronic*; 2009. Available from: https://www.ipc.on.ca/images/resources/privacybydesign.pdf. (Visited on 04/04/2016).

[15] Notario Nea. "PRIPARE: integrating privacy best practices into a privacy engineering methodology". In: *Security and Privacy Workshops (SPW), 2015 IEEE*. Piscataway, NJ: IEEE; 2015. p. 151–158.

[16] Deng M, Wuyts K, Scandariato R, Preneel B, Joosen W. "A privacy threat analysis framework: supporting the elicitation and fulfillment of privacy requirements". *Requirements Engineering*. 2011;16(1):3–32.

[17] Hoepman JH. "Privacy design strategies". In: *ICT Systems Security and Privacy Protection*. Berlin: Springer; 2014. p. 446–459.

[18] Key Steps for Organizations in Responding to Privacy Breaches. Available from: https://www.priv.gc.ca/information/guide/2007/gl_070801_02_e.asp. (Visited on 04/04/2016).

[19] Perez JC. "Facebook's beacon more intrusive than previously thought". *Electronic*; 2008. Available from: http://www.pcworld.com/article/140182/article.html. (Visited on 04/04/2016).

[20] Zuckerberg M. "Thoughts on beacon". *Electronic*; 2007. Available from: https://www.facebook.com/notes/facebook/thoughts-on-beacon/7584397130. (Visited on 04/04/2016).

[21] Mathews AW, Yadron D. "Health insurer anthem hit by hackers". *The Wall Street Journal*; 2015. Available from: http://www.wsj.com/articles/health-insurer-anthem-hit-by-hackers-1423103720. (Visited on 04/04/2016).

[22] Gross D. "Google Buzz goes after Facebook, Twitter". *Electronic*; 2010. Available from: http://edition.cnn.com/2010/TECH/02/09/google.social/. (Visited on 04/04/2016).

[23]   Wikipedia. "Google Buzz". *Electronic*; 2016. Available from: https://en. wikipedia.org/wiki/Google_Buzz. (Visited on 04/04/2016).

[24]   LLC PI. *2015 Cost of Data Breach Study: Global Analysis*. Traverse City, MI: Ponemon Institute LLC; 2015. Available from: https://security intelligence.com/cost-of-a-data-breach-2015/. (Visited on 04/04/2016).

[25]   Cornish A. "How much will the hack cost Sony?". *Electronic, Audio*; 2014. Available from: http://www.npr.org/2014/12/18/371721061/how-much-will-the-hack-cost-sony. (Visited on 04/04/2016).

[26]   Sony Corporation. "Consolidated financial results forecast for the third quarter". *Electronic*; 2015. Available from: http://www.sony.net/SonyInfo/IR/library/fr/150204_sony.pdf. (Visited on 04/04/2016).

[27]   Dean B. "Why companies have little incentive to invest in cybersecurity". *Electronic*; 2015. Available from: http://theconversation.com/why-companies-have-little-incentive-to-invest-in-cybersecurity-37570. (Visited on 04/04/2016).

[28]   Weber RH. "Internet of Things – need for a new legal environment?" *Computer Law & Security Review*. 2009;25(6):522–527.

[29]   Solove DJ. "'I've got nothing to hide' and other misunderstandings of privacy". *San Diego Law Review*. 2007;44:745.

[30]   Rubinstein IS, Good N. "Privacy by design: a counterfactual analysis of Google and Facebook privacy incidents". *Berkeley Tech LJ*. 2013;28:1333.

[31]   National Institute of Standards and Technology (NIST). *Standard 8062 – Privacy Risk Management for Federal Information Systems*. Gaithersburg, 2015.

[32]   Spiekermann S, Cranor LF. "Engineering privacy". *Software Engineering, IEEE Transactions on*. 2009;35(1):67–82.

[33]   Internet Privacy Engineering Network. "IPEN live". *Electronic*; 2014. Available from: https://secure.edps.europa.eu/EDPSWEB/edps/EDPS/IPEN. (Visited on 04/04/2016).

[34]   Koops BJ. "The trouble with European data protection law". *International Data Privacy Law*. 2014;4(4):250–261.

[35]   European Commission. "Why do we need an EU data protection reform?". *Electronic*; 2012. Available from: http://ec.europa.eu/justice/data-protection/document/review2012/factsheets/1_en.pdf. (Visited on 04/04/2016).

[36]   Cavoukian A. *Privacy by Design: The 7 Foundational Principles*. Ontario, Canada: Information and Privacy Commissioner of Ontario, Canada. 2009.

[37]   Gürses S, Troncoso C, Diaz C. "Engineering privacy by design". *Computers, Privacy & Data Protection*. 2011;14.

[38]   García AC, McDonnell NN, Troncoso C, *et al. Privacy- and Security-by-Design Methodology Handbook*. PRIPARE Project; 2015. Available from: http://pripareproject.eu/wp-content/uploads/2013/11/PRIPARE-Methodology-Handbook-Final-Feb-24-2016.pdf.

[39]   Corporation M. *Security Development Lifecycle – SDL Process Guidance Version 5.2*; 2012. Available from: http://www.microsoft.com/en-us/download/confirmation.aspx?id=29884. (Visited on 04/04/2016).

[40]   Kalloniatis C, Kavakli E, Gritzalis S. "Addressing privacy requirements in system design: the PriS method". *Requirements Engineering*. 2008;13(3): 241–255.

[41]   Howard M, Lipner S. *The Security Development Lifecycle: SDL: A Process for Developing Demonstrably More Secure Software*. Redmond, WA: Microsoft Press; 2006.

[42]   Solove DJ. "A taxonomy of privacy". *University of Pennsylvania Law Review*. 2006;154(3):477–564, 2006.

[43]   Doty N, Zych J, Gupta M, McDonald R. "Privacy pattdata database". *Electronic*; 2011. Available from: http://privacypatterns.org. (Visited on 04/04/2016).

[44]   Willis LE. "Why not privacy by default". *Berkeley Tech LJ*. 2014;29:61.

[45]   Unabhängiges Landeszentrum für Datenschutz. "Seal of privacy for IT products and privacy protection audit for public authorities". *Electronic*; 2002. Available from: https://www.datenschutzzentrum.de/uploads/ guetesiegel/Seal-of-privacy-and-privacy-protection-audit.pdf.   (Visited   on 04/04/2016).

[46]   Zimmermann C, Accorsi R, Muller G. "Privacy dashboards: reconciling data-driven business models and privacy". In: *Availability, Reliability and Security (ARES), 2014 Ninth International Conference on*. Piscataway, NJ: IEEE; 2014. p. 152–157.

[47]   Cuellar J, Suppan S, Waidelich F, *et al. System Requirements and Smart Objects Model*. RERUM Project; 2014. Available from: https://bscw.ict-rerum.eu/pub/bscw.cgi/d14593/RERUM%20deliverable%20D2_2.pdf. (Visited on 04/04/2016).

[48]   RERUM. *Deliverable 2.3 – System Architecture*. RERUM Project; 2014. Available   from: https://bscw.ictrerum.eu/pub/bscw.cgi/d14593/RERUM% 20deliverable%20D2_2.pdf. (Visited on 04/04/2016).

[49]   Bassi A, Bauer M, Fiedler M, *et al.*, editors. *Enabling Things to Talk: Designing IoT Solutions with the IoT Architectural Reference Model*. Berlin: Springer; 2013.

[50]   Pearson S, Mont MC. "Sticky policies: an approach for managing privacy across multiple parties". *Computer*. 2011;44(9):60–68.

[51]   Mont MC, Pearson S, Bramhall P. "Towards accountable management of identity and privacy: sticky policies and enforceable tracing services". In: *Database and Expert Systems Applications, 2003. Proceedings. 14th International Workshop on*. Piscataway, NJ: IEEE; 2003. p. 377–382.

[52]   Karjoth G, Schunter M, Waidner M. "Privacy-enabled management of customer data". *IEEE Data Engineering Bulletin*. 2004;27(1):3–9.

[53]   Suppan S, Cuellar J. "Datenschutzbewahrende Policies für die Kommunikation im Internet der Dinge". *Prior Art Journal*. 2015;4.

[54]   Manulis M, Fleischhacker N, Gunther F, Kiefer F, Poettering B. "Group signatures: authentication with privacy". *Bundesamt für Sicherheit in der Informationstechnik (BSI)*; 2012. Available from: http://www.bsi.bund.de. (Visited on 04/04/2016).

[55]   Rannenberg K, Camenisch J, Sabouri A. *Attribute-Based Credentials for Trust: Identity in the Information Society*. Berlin: Springer; 2014.

[56]   Titkov L, Poslad S, Tan JJ. "An integrated approach to user-centered privacy for mobile information services". *Applied Artificial Intelligence*. 2006;20 (2–4):159–178.

[57]   Scheuer F, Posse K, Federrath H. "Preventing profile generation in vehicular networks". In: *Networking and Communications, 2008. WIMOB'08. IEEE International Conference on Wireless and Mobile Computing*. Piscataway, NJ: IEEE; 2008. p. 520–525.

[58]   Bormann C, Ersue M, Keranen A. *Terminology for Constrained-Node Networks [RFC 7228]*; *Internet Engineering Task Force (IETF)*, 2014 Available from: *https://tools.ietf.org/html/rfc7228*.

[59]   Memory Protection Devices. "CR2032 datasheet and articles for designing with CR2032 batteries". *Electronic*; 2011. Available from: http://www.cr2032.co/datasheets.php. (Visited on 04/04/2016).

[60]   Gerdes S, Bergmann O, Bormann C. "Delegated CoAP authentication and authorization framework (DCAF)". IETF draftgerdes-core-dcaf-authorize-02; 2014.

[61]   Cuellar J, Suppan S, Poehls HC. "Privacy-enhanced tokens for authorization in ACE" draft-cuellar-ace-pat-priv-enhanced-authz-tokens-02; 2016.

[62]   Yan Z, Zhang P, Vasilakos AV. "A survey on trust management for Internet of Things". *Journal of Network and Computer Applications*. 2014;42:120–134.

[63]   Gambetta D. "Can we trust trust?" *Trust: Making and Breaking Cooperative Relations*. 2000;13:213–237.

[64]   Kagal L, Finin T, Joshi A. "Trust-based security in pervasive computing environments". *Computer*. 2001;34(12):154–157.

Chapter 3
Privacy and consumer IoT: a sensemaking perspective

Gaurav Gupta

Summary

Internet of Things (IoT) is slowly entering the realm of consumer devices in the form of smart devices and home automation. However, simultaneously it has led to concerns of privacy due to the frequent citation of media reports highlighting weak privacy policies of device manufacturers [1]. In some cases, manufacturers have even been found themselves actively monitoring and analyzing user's private information [2]. End user's private information is exchanged and traded as currency in this IoT ecosystem. These apprehensions have affected consumer IoT adoption. In this chapter, novel concerns of consumer privacy are highlighted. The change in connected device architecture is presented, to reflect upon the role of various stakeholders in this ecosystem. Sensemaking theory is used to explain the change in the relation between users and conventional consumer devices vs IoT-enabled devices. Based on this renewed understanding, we engage the privacy concerns and system use perspectives of IoT users and providers focusing on enhancing perceived usefulness of IoT in consumer devices. Future avenues for research are suggested, focusing on designing privacy-aware systems.

3.1 Introduction

Adoption of updated consumer technology is non-optional for end users. Older technologies are replaced by newer ones thereby restricting access to older devices and technologies. Internet of Things (IoT) was discussed as a conceptual model by Kevin Ashton in 1999 to connect physical objects from our real world with the Internet in the context of supply chain management [3]. Since then, the technology has matured and is now implemented in consumer devices. Consumer appliances majors like Samsung have shifted their focus to IoT device markets and are integrating this technology into their existing product offerings [4]. The immense potential offered by the possibility of collection of data from devices that are an irreplaceable part of our routine lives has also encouraged advertisement revenue-dependent organizations to offer IoT services.

Recently, Google too unveiled its IoT offerings Project Brillo and Google Weave to extend its smartphone platform (i.e. Android) to low-power IoT devices [5].

IoT is no more a theoretical concept but a practical application of ICT that cohabitates with us in our living world. Under IoT framework, the everyday devices that we use connect to various other devices and servers to collect and process information, as diverse as our behavioral preferences to our body parameters unceasingly, with different agencies in this ecosystem. Information privacy is defined as "the ability of the individual to control the terms under which personal information is acquired and used" [6]. Legislation is one method of ensuring privacy in connected environments [12] but identification of jurisdiction is a prerequisite of any legislation. In the case of IoT, information ownership is non-singular and does not remain static with any one entity and traverses among many actors, with multiple manipulations of the input information at each stage. Assigning ownership and responsibility for ensuring the privacy of user information hence becomes difficult. Therefore, legislation will be difficult to implement without a clear understanding of these underlying issues. Ownership and intended usage of both primary information (collected from individuals with or without their explicit knowledge) and auxiliary information (conceived from analysis of primary information) are essential for information privacy [7,8].

This chapter engages with the user concerns of privacy in IoT and then contrasts it with other conventional connected architectures. A set of propositions are developed to understand the privacy matrix encompassing the users and the sources of diffusion of provided information, from the perspective of changes in the social order due to the introduction of IoT. This understanding will help industry practitioners, IoT standards drafting agencies, policy makers, IoT device manufacturers and other IoT infrastructure providers to realize the concerns of information ownership when designing a robust privacy-aware IoT framework by consolidating these different perspectives. With concerns about privacy managed adequately, acceptance, adoption and diffusion of IoT devices in user's routine shall improve too.

The chapter is structured as follows. The next section discusses the novel concerns of information privacy both from the perspective of end users and IoT device providers. With these concerns highlighted, the next section presents the change introduced in constructs of ownership and control of information due to IoT as a technology in routine consumer appliances. An adaptation of sensemaking theory is then proposed to fill the gap between user's and provider's aspiration of control of information in this ecosystem. The chapter concludes with a discussion on the future direction of research especially focusing on suggestions to calibrate these theoretical insights to application design.

3.2   Consumer IoT and novel privacy concerns

Novel capabilities conferred to consumer devices by IoT framework requires identification of security and privacy concerns of various stakeholders. Fleisch [9] identified several value drivers for IoT applications. Present IoT consumer applications present at least one or more of these value propositions. All of these value drivers could be categorized into application for automation of some form of user stimuli for process initiation or assistance in providing feedback for process progression. IoT bridges the gap between physical world and the digital world. Gershenfeld [10] described IoT as an extension of the Internet to reach out to the physical world of things and places that only can feature low-end computers. Integration of physical objects from our social surroundings with the open Internet boosts apprehensions related to information privacy too.

Consumers provide personal information to IoT systems in order to automate repetitive stimuli. Apart from that, information of the operating environment and user preferences is also collected by these devices (seldom without user's knowledge). Both these forms of collected information can be further analyzed together to capture greater insights into consumer behavior and preferences. To achieve acceptability by consumers, IoT deployments must possess the properties of openness, viability, reliability, scalability, interoperability, data management and security solutions [11]. Privacy concerns of IoT device providers are rooted in their intention to ensure user's trust on the infrastructure, resulting in gradual adoption.

From the perspective of IoT device manufacturers, consumer privacy concerns can be categorized into the intentional and unintentional diffusion of information by themselves and by other agencies forming the whole IoT infrastructure. Unintentional diffusion of private information can be attributed to malicious actors who access private data during transmission, storage or retrieval in the system. Eavesdropping, IoT node capture and controlling of IoT infrastructure are some of the techniques employed by such actors [11]. Multiple Privacy Enhancing Technologies (PET) have been designed to deter and resist such attempts. Techniques like Virtual Private Networks, Transport Layer Security, DNS Security Extensions, Onion Routing, Private Information Retrieval systems and Peer-to-peer systems help deter such threats [12]. However, deterrence does not ensure complete compliance. IoT framework is designed by governing bodies like IPSO, IIC, etc. which ensure adherence to privacy protocols. However, heterogeneity in standards and architecture of IoT aggravates the complexity of security threats and privacy concerns [13].

Intentional diffusion of consumer information transpires when manufacturers share data with other agencies for marketing and other purposes. This leads to suspicion and distrust in the minds of the customer. Radio Frequency Identification (RFID) along with Object Naming Service (ONS) form the backbone of IoT implementation. European Commission in a Recommendation of May 12, 2009 [14], suggested provisions for transparency on RFID use, the RFID applications used in the retail trade, the awareness raising actions, research and development as well as follow-up actions (Nos. 7–18). Weber [12] suggested legal recourse along with technological support to enable privacy-aware RFID-based framework design. The non-existence of universal laws and incompatibility of state laws to ensure transparency of data usage collected by IoT devices are serious concerns in ensuring privacy to the end users.

In an online environment, consumer concerns about privacy span his intent of controlling the actions of the device manufacturer and associated agencies and restricting the secondary use of provided information [15]. Similarly, owing to connected nature of IoT devices, consumer concerns of IoT-enabled devices are primarily associated with control of the device's action in relation to its communication with the back-end server and secondary usage of aggregated information. Loss of perceived control over private information shared with the device leads to reduced acceptance during implementation phase [16]. As illustrated in the next section, the user might own the device but its functionality is supported by a technology back end on which he might not have any control. To automate and improve its offerings, IoT devices collect data actively and unceasingly, even when switched to hibernation mode. End users exercise minimal authority on access control to this information and hence are dependent on IoT providers to secure it.

For non-IoT devices, the end user was the customer for device manufacturers. Value capture from the end user was the driver for their business activities. Therefore, customers were given complete control of customization of their user experience. They were given the choice to use them with limited functionality if needed. A television set could be used as an output display without connecting them to the satellite network, and it would still work. However, IoT-devices do not serve the end users solely. They have other value-generating customers too in the form of marketers. For providers of IoT infrastructure (including manufacturers), primary value is derived from device adoption and usage by end customers and their respective behavior analysis. This data provides direct and indirect revenue to the IoT device manufacturers and hence are collected simultaneously while providing primary functionality. Manufacturers may themselves analyze user behavior data to improve or automate their product offerings. They may also further share this data with other external agencies to gain further insights on their behavior or even earn revenue from this transaction. Due to this segregation of value drivers, IoT providers acquire supplementary user data if the value obtained out of it can be captured by them with minimal costs. Hence, these devices are designed to become nonfunctional without connection to back-end IoT infrastructure.

Furthermore, unlike non-IoT devices, detachment of these devices from the network at a later stage in order to safeguard private information is arduous for the user. Communication capabilities of non-IoT devices could easily be curbed by disconnecting them from the network. However, disabling of tracking by RFID-based IoT devices can only be accomplished by putting them in a Faraday Cage or by deleting information from ONS [12], both of which are difficult to execute for most users. Moreover, poor system design may make these devices unusable for their primary functionality when disconnected from the IoT infrastructure.

Our deep-seated adaptation of these consumer devices to our routine lives owes to their irreplaceability from our social spaces. Hence, these privacy concerns with our routine devices necessitate scrutiny of the underlying privacy framework pre- and post-implementation of IoT as a feature.

3.3   IoT-induced change in information architecture of consumer devices

Consumer appliances have become an irreplaceable part of our routine lives. Before the advent of consumer IoT devices, the end user owned the device and also all

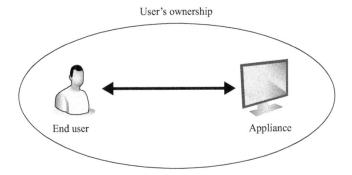

Figure 3.1   General framework for information ownership pre-IoT

the input information. His ownership of the device manifested into control of the associated information. For example, a television set received information about user's watching habits, etc., but the television manufacturer or any other agency associated with the device had no access to that information about his preferences. Any information collected resided on the device itself and was not shared back with the manufacturer or any associated agencies.

As represented in Figure 3.1, the feedback loop for any information supplied to a non-IoT device (e.g. user preferences) encompassed just the user and the device itself. Although many of these consumer devices did connect to external networks (like satellite TV) to fetch content requested by the user, they did not share any information back with the source. Any other agency did not have access to this information and, therefore, the user's data was kept private. With this understanding of devices existing around them, the users allowed its presence to fade into the background of their routine social lives. Since the assumption was that these devices did not communicate with external data assimilating agents, users with varying privacy concerns shared their living spaces with those devices unreservedly and thus made them witness their routine private lives.

For consumer IoT devices, the concepts of ownership and control are murky. Sole proprietorship of information is unheard of. The connection with the information source is bidirectional. For example, an IoT-enabled refrigerator communicates back and forth with the server to retrieve the user expected information when requested. It accepts user's preferences and connects with multiple web servers, as needed, to process and retrieve desired response. A sample use case is of an IoT-enabled refrigerator ordering food items if the supply inside falls below a threshold. It stores user-desired inventory of food items and engages in two-way communication with e-grocery platforms to order supplies. Rodrigo [11] proposed four models of IoT architectures: centralized, collaborative, connected and distributed. Distributed control of data is obligatory in all of these models. None of the present models offer user proprietorship of information.

The user owns the device, but its IoT-backed functionality is provided as a service and hence dependent on the communication with the external server. Users would expect network independent basic functionalities of the device will not require it to connect back to the servers, but this is hardly the case. Even to provide basic functionalities, IoT devices tend to be dependent on the IoT infrastructure.

The user does not have complete control of all the information collected by the device. The typical case is of a user requesting some information from his device, and this request been collected and sent to the manufacturer's server for processing. Once a request has been sent it can travel in either of the following paths – the manufacturer's server may process it and respond directly back to the user, or it may send that request to an external server for processing that in turn responds back with the processed information to be sent to the user. In the first case, the user sends his information to the manufacturer's server where the request is processed and fulfilled completely by its servers only. For example, a user's query about servicing information of the IoT device would send device and owner info to the manufacturer's server and respond appropriately without needing the intervention of any other agencies. User's information is shared with the manufacturer who might save or further analyze this to extract additional information about the user's behavior for other undeclared purposes. The device manufacturer might itself share this information with other agencies for profiling using this data for purposes like advertising.

If the request cannot be completely fulfilled by the manufacturer's server solely, then it will be sent in full or in parts to some external web servers for processing. The earlier example of the fulfillment of the request for restocking of an IoT-based refrigerator will need it to interact with multiple other web servers like payment gateways, e-commerce platforms, etc. Both the manufacturer and the independent web server owner can save and further process the user's information to get greater insight into the user's behavior or to predict his actions. Each of these actors may share this information with other agencies too, and this seldom happens oblivious to the user's knowledge.

In both of these cases, the user owns the IoT device but he does not have complete control of shared information. Figure 3.2 represents the conceptualization of ownership of different actors in IoT-enabled consumer devices. An IoT-enabled device's morphology extends beyond the user-owned hardware. Since these devices strive to process user-generated information requests by connecting to servers, the expanse of user ownership of shared information does not completely encompass the device's physical form. Since all the information requests originating from the device has to be supplied to the manufacturer's server for processing and the response is also controlled by it, the device cannot provide all its functionality independently. Thus, the device manufacturer too shares some control on the device. Apart from that, often the manufacturer is unable to process these requests independently and so sends a request for further processing to other external servers too. These external agencies can collate, save and reprocess this information further. The manufacturer has no control over the information shared with these external agencies. Also due to vulnerabilities in the security architecture of IoT devices, they sometimes expose user data to data theft [17]. Therefore, at each stage of this information processing, the control of user's behavioral information is shared with multiple agents although the ownership of the IoT device resides with the user.

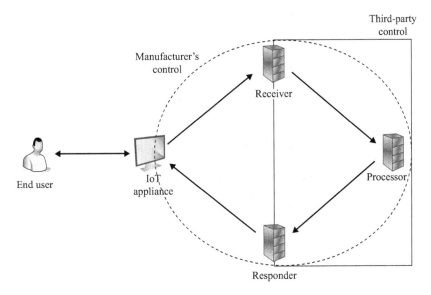

Figure 3.2   Information ownership framework for IoT devices

The purpose of IoT devices is to automate repetitive tasks by collecting informa-tion from human interactions [18]. Therefore, apart from explicit queries by end users, these IoT devices unprovokedly also do collect much information from their array of sensors to preprocess expected information requests. The collected information can encompass both the owner and the environment (usually in the device's vicinity). The extent of this information collection is not disclosed to the end user usually, and thus it makes their activities suspicious. Even if it is disclosed, it is made imperceptible by veiling it behind requests for information to perform the primary functions of the device.

Since the conventional norms of ownership and control are modified in the con-text of household IoT-enabled devices, a new set of rules would have to be defined to explain these changes. To make sense of these new rules and thus to establish it, we would have to unlearn and then relearn our understanding of privacy in the context of newly found communication capabilities in our conventional consumer devices to allow their coexistence in our social life.

3.4   Sensemaking of information privacy and IoT

Sensemaking, as the term suggests, refers to the process of making sense of a situation or condition. It is a process of organization, selection and categorization of all per-ceptions to aid in the interpretation of events, objects and people [19]. Sensemaking is viewed as both an individual and a social activity by acknowledging the impact of others on the sensemaker [20]. Implicitly, it also suggests discontinuity from the adaptive structuring of realized understanding over time and creation of new understanding. Sensemaking has been adopted extensively for understanding the process of strategic change initiation in organizational culture or practices [21–23]. Weick [20] presented seven properties of sensemaking in organizations: retrospective, grounded in identity construction, enactive of sensible environments, social, ongoing, focused on and by extracted cues and driven by plausibility rather than accuracy. All of these need to be assimilated for the recipe of change to be implemented in an organization. Management literature has used sensemaking for explaining the context of new technology in organizations [24], reading policy by teachers in professional communities [25], organizational restructuring [26], creativity in organizations [27] and so on.

Information privacy and monitoring are closely related concepts. While the first refers to an individual's right to let be, the other attempts to curb it. Technology has enabled rich, pervasive and intensive monitoring without the need for a physical supervisor, thus making surveillance innocuous. Alge *et al.* [28] used the concept of sensemaking to study the phenomenon of information privacy and organizational monitoring with the lenses of identity and fairness. He argues that knowledge of monitoring by a technological system acts as an attentional cue for individuals, causing one to be highly self-aware and concerned with one's own personal and social identities resulting in a sense of discontinuity. Individuals attempt to make sense of this monitoring by assessing the system's social and symbolic meaning beyond the apparent technology features of the system [29]. These discontinuities allude the individual to reestablish their privacy stance with respect to perceived surveillance.

Sensemaking is a multistage process that is enacted to understand a crisis situation [30]. It starts with the realization of the present situation and identifying how the environment and the situation are enacted by the people. Sense breaking marks the beginning of this process of relearning. It begins with the collapse of embeddedness of identities and beliefs in the present situation. Our understanding of our environment is a result of both our logical and normative understanding. We construct it on some basal facts that lead to enacted consequences of understanding. Understanding is also created as a result of preconceived norms and beliefs that are then conceived as facts. The collapse of this understanding creates a void for rethinking and reconceptualization of those. Not all understanding is corroded, but the ones directly related to the creation of logical or normative knowledge is lost. For example, in crisis situations constructs like trust are mutilated. People lose their trust in processes, and people start questioning the basis of their trust. This leads to corrosion of normative beliefs and spawns a void for the creation of new understanding (see Figure 3.3).

However, once an accepted sense of understanding is lost, then this abandonment leads to demanding of novel understanding of the present situation. Sense demanding sets the stage for acceptance of new norms, beliefs and identities. Constructs like trust are adapted for rebuilding. Due to the collapse of understanding, vulnerability to different agents of sense giving is heightened. Simultaneously, the creation of new understanding also becomes difficult due to lack of normative and basal understanding. Sense demanding is characterized with a cautious eagerness to redefine the collapsed constructs, with heightened awareness of the present situation or condition.

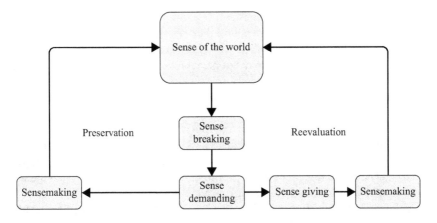

Figure 3.3   Sensemaking evolution

Sense giving by multiple agents leads to initiation of the process of creation of new understanding. Unlike acquired understanding that is a mix of normative and logical knowledge, this process is highlighted by the acquisition of factual knowledge. The void of old constructs is repaired to create new rules of understanding. Multiple agents co-create new structures of understanding for readoption.

The context of IoT-backed capabilities in our consumer devices forces us to re-evaluate our relationships with these devices. They did exist in our routine social spaces since years (without IoT features), and users were aware of their capabilities and could control their actions. However, as discussed in the previous section, users can own IoT devices but they cannot completely control the flow of information through them. Ownership of the devices' actions is shared between the user and other agencies. This creates a sense of discomfort and distrust due to the presence of these devices in our social spaces. To reestablish familiarity in the relation between conventional consumer devices and their users, rethinking and reevaluation of their stature in our spaces is necessary.

The source of user's discomfort is not the whole device but rather some of its capabilities. These devices had been cohabiting in our social spaces for quite some time and had faded their presence into the subconscious. They had become infinitesimal artifacts in our routine social lives. However, the emergence of IoT capabilities had made them reappear in our conscious psyche. Although users want to continue using the device as ever but they feel uncomfortable of IoT-backed features, especially those related to communication capabilities of the device. Desanctis and Poole [31] separated technology structure into its features and intent. Intent refers to the base functionality of the technological device and features refer to the decisions made during design time. The device can be conceptualized by manipulating some of its features, but its intent is clearly defined and rigid. Design intent and use intent should match exactly in order to realize a functional device. Griffith [32] proposed a conceptual model of technology characteristics and its features. With this model, he argued that a significant gap exists between design-driven and user-driven technology features. Many of these features are not used or even their presence realized by most users and hence they remain dormant. However, even their dormant presence gives the device some capabilities that may not be approved by the user. Often the design features and use features do not perfectly align, which leads to discord during its adoption. Many features integrated into the product are sometimes not even noticeable by the users and the ones that are noticed are adapted over a period by a process of social construction. Scholars like MacMillan and McGrath [33] noted the cruciality of the ability to identify the features that the users will find significant and the ones that may be disastrous to the product (due to its social construction and deconstruction by the users). The adapted form of these functionalities may be different from the ones provided by design. This dissonance affects adoption intent of the device.

In this context, consumer devices can be explained in terms of their features and intent, pre- and post-IoT enablement. Prior to IoT implementation in consumer devices, their intent was clear. A television set was expected to display satellite network beamed and thus present the requisite information. Similarly, a refrigerator was only expected to keep eatables cold and fresh. These devices had some extra features built into them like a wireless remote control for televisions and ozonizing capabilities in refrigerators. These features aimed to improve the usability of these devices but did not interfere with the basic intent. Some of these devices had communication capabilities as some of these did connect to satellites etc. to fetch information. Therefore, in terms of intent pre-IoT devices were not very different from IoT-enabled ones. The difference lies in the way these features and the overall intent of these devices is modified by the introduction of IoT. The intent of these devices is bifurcated i.e. primary intent and secondary intent. The primary intent of these consumer devices still remains the same, but a secondary intent of these devices has developed, which is to automate repetitive tasks. For example, an IoT-enabled refrigerator apart from cooling its contents might also be capable of executing repetitive tasks like ordering grocery when supplies fall below a threshold. An IoT-enabled television is also capable of displaying pre-fetched contextual information along when requested. IoT infuses them with partial autonomy to execute many of its routine activities. Automation requires data capture that might or might not have been exclusively made available to the device. As newer features as capabilities are introduced to the device, these data capture requirements do increase too and simultaneously our discomfort with their optimal usage. Sensemaking is triggered hence, causing establishment of new identities for these devices.

Feature-based theory of sensemaking is proposed to fill this gap in technology design and its perceived usefulness. It helps link extant understanding of sensemaking triggers with later stage models of technology understanding and use (like adaptive structuration theory) [32]. Triggers like suspicion of communication capabilities of IoT devices are apt for the application of feature-based sensemaking theory. However, it is applied in the technology introduction phase [32], unlike consumer IoT devices that have already existed and adapted to our social spaces with their primary functionalities (intent). However, since their novel capabilities, introduced by IoT, has reconceptualized the overall intent of the device, therefore the devices can be considered to have been relegated to the introduction phase again as a new device. Hence, feature-based sensemaking theory can be applied in the context of re-sensemaking of existing consumer devices triggered by novel capabilities introduced into them.

According to Griffith's classification of technology features [32], IoT-backed features are core rather than tangential since their presence is acknowledged by users during usage of the device for its primary intent. It is impossible to use these devices without interacting with IoT-backed features. Also, IoT-backed features form the core intent of the device. Their presence can be ascribed to verifiable fact, and it is not possible to discount their presence. Interaction with the device is impossible without acknowledgment of its IoT capabilities. Hence, IoT features are both core and concrete. According to feature-based sensemaking theory, these new IoT-backed features are more likely to be experienced as novel and discrepant owing to their concreteness and core characteristics. These features, therefore, trigger sensemaking leading to adaptive structuration and later regelation again into our routine social spaces. However, it would need conceptualization of the relationship between producers and diffusers of user's private information regarding their concerns and needs of control of this data.

3.5 Integration of information privacy needs and concerns in IoT

The context of triggering of sensemaking due to IoT according to feature-based sense-making theory, as discussed in the previous section, is different from the introduction of new technology. Adoption theories like Technology acceptance model (TAM) have been used by information systems researchers to explain the process of behavioral intention to use a new technology [34–37]. However, they are not designed to explain the phenomenon of reintroduction (with changes in the core intent) of an existing technology when the user has prior exposure to a similar technology. Constructs like experience moderate the interaction of adjustments to the technology with user's perceived ease of use, before direct exposure to the technology. Griffith's model [32], which distinguishes a technology from its features, helps explain the case of the reintroduction of consumer appliances with IoT capabilities. Accordingly, the technological device (i.e. consumer appliance) with its primary intent was exposed to the users already, and it had diffused into their routine. However, enhancement of its primary intent with IoT capabilities led to a sense of distrust on these consumer devices. Hence, their intention to use these devices, along with IoT-backed features, again will depend on multiple factors.

Marketing literature has recognized that significant gaps do exist between consumers' expectation of privacy in connected environments against marketers' expectations [38]. With the following set of propositions, this study proposes to combine user's information privacy concerns and IoT provider's information assertion in the context of perceived control over information. This study also attempts to explain how these affects usage intention, moderated by legislative controls.

3.5.1   User's privacy concerns and perceived control of information

Private information can be distinguished from one that leads to embarrassment or one that leads to loss of privacy or control [39]. Disclosure about the collection of information and transparency of its usage will allow users to make an informed decision that improves their trust on the ecosystem.

People are averse to sharing some kinds of information like medical, financial of family related but far less uncomfortable sharing certain others like product or brand preferences. According to Westin's privacy segmentation, individuals too differ in their concerns about information privacy ranging from "privacy unconcerned" to "privacy pragmatists" to "privacy fundamentalists" [40]. The basis of their classification into this schema is based on their perception of loss of control over how personal information is collected and used by companies. In the context of IoT, these companies consist of multiple agencies like IoT device manufacturers, network providers, third-party information processors, etc. Users who belong to the category of privacy fundamentalists do firmly believe in their loss of control over personal information and its subsequent secondary usage. Privacy pragmatists will relinquish significant control of their private information and are willing to trade it for perceived benefits. Against this, privacy unconcerned users may not believe in the said loss of control or are not bothered even if it is true.

Proposition 3.1. *Users with greater concerns about privacy will seek to achieve greater perceptive control of the process of collection and usage of their private information.*

3.5.2   Provider's information assertion and perceived control of information

IoT device manufacturers attempt to enrich devices with more features to enable automation of user stimuli at the expense of weak privacy architecture. The absence of avenues for face-to-face communication requires firms to use consumer information to offer personalized services that will increase value and hence consumer loyalty [41]. The value generated by these activities may not necessarily be the value demanded by consumers but further leads to increased system complexity and usage anxiety. Venkatesh [42] argued that computer anxiety was one of the precedents to perceived ease of use in TAM. Users do vary in their anxiety in using these devices. So manufacturers regulate default usage behavior of these devices for users to minimize conscious deliberation and associated usage anxiety.

Apart from that, some IoT providers attempt to capture business value by sharing consumer's private information with external agencies for behavior analysis. Such providers tend to maintain information opacity about device behavior as knowledge of such activities will dissuade privacy-concerned users. Therefore, such firms tend to indulge in these activities covertly by regulating default usage. Zuboff [43] asserts the existence of a market for trade of consumer's private information. She termed it as surveillance capitalism. Firms vary in their assertion of user-shared information although providing the same services due to the distinction of their business objectives.

Proposition 3.2. *Firms with greater assertion on user's information will allow lesser control of their private information.*

3.5.3 Legislative framework and perceived control of information

International IoT policy regulators have been invoked to engage in the debate of allowing individuals the ability to disconnect from the network anytime and putting in place necessary regulations to ensure it [14]. Weber [12] suggested setting up of a legal framework to ensure compliance of consumer privacy regulations, apart from self-regulation by firms themselves. It would ensure both deterrence and compliance to agreed privacy norms. Since IoT infrastructures extend globally through a network of servers and protocols, ensuring compatibility with these regulations would pose a significant barrier to realizing this vision. Individual states would have to ensure that their regulations comply with each other. Hence, Weber [12] suggested the establishment of a global legislator for policy making and ensuring applicability of its regulations concerning privacy norms to all objects (including information) from their genesis to dissolution. Adherence to its regulated privacy framework would ensure greater control of private information in the ambit of consumers.

Proposition 3.3. *Greater robustness of privacy ensuring regulations would lead to greater control of information for the end users.*

3.5.4 Perceived control of information and intention to use

Information systems and psychology literature have treated control as a construct to explain user's intention to act. It specifically relates to Bandura's explanation of perceived self-efficacy that "is concerned with judgments of how well one can execute courses of action required to deal with prospective situations" [44]. Ajzen and Fishbein [45] explained behavior and intentions by Theory of Planned Behavior (TPB) which extended the Theory of Reasoned Action [46,47] by primarily adding the perception of behavioral control. They argued that perceived control of one's action toward the achievement of expected outcomes, influences intention to act and its ultimate successful conclusion. This conceptualization of control refers to both internal and external control. Internal control refers to knowledge and self-efficacy while external control refers to the environment [48].

In the context of IoT, the user's perception of control over his private information influences his intention to share it with the IoT device (assuming it to be the prerequisite for its usage). His perceptions of control extend to control of the process of information collection (i.e. knowledge and self-efficacy) and control of usage of provided information (i.e. control on environment).

Proposition 3.4. *Greater the amount of perceived control (both internal and external) a user exercises over his private information, more will be his intention to use the system.*

Figure 3.4 illustrates propositions relating user's privacy concerns with IoT provider's information assertiveness and legislation capabilities, with perceived

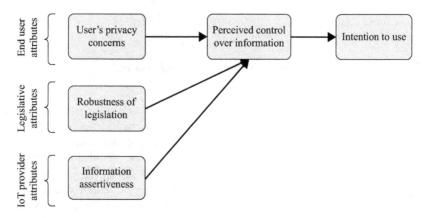

Figure 3.4   Characteristics, control and intention

control of information and later adoption. In summary, behavioral intention to use IoT system shall improve by allowing greater control of information to the end customer. Improving the legislative structure to ensure that even if firms may be seeking to assert greater control over user information, will allow privacy conscious users to intend to use these devices more.

3.6   Discussion

Sensemaking theory provides a firm outlook to the phenomenon of introduction of IoT features in consumer appliances. As mentioned previously, adoption theories in general do not consider the scenario of prior acquaintance with similar technologies. But sensemaking theory acknowledges this discontinuity and further implores the reconditioning of accepted constructs like perceived usefulness and perceived ease of use. Moreover, it enlightens the IoT providers to understand the expectations of the consumers and simultaneously helps the consumer to understand the logic of automation. This abridgment shall lead to quicker readoption of these feature laden appliances.

In line with earlier research that consumers should be given greater control during information collection and dissemination [49], this study supports assigning control of private information to the end user. IoT adoption can be improved by IoT providers by allowing control of both the process of collection and usage of private information. But granting micro-level control of the device features to the end user might cause user anxiety and hamper adoption. Similarly, high level of automation might cause discomfort for the user on grounds of disempowerment of control rights on the device. Allowing optionality to collection of noncritical data can be one way of impressing user control of information. IoT devices should provide the ability to disable IoT-backed features of the device and use it for only its primary intent. Therefore, users with varying levels of privacy concerns will be catered to and hence improve acceptability of IoT in consumer devices.

Drafting of an appropriate framework for management of globally compatible privacy controls will also help increase user trust in the ecosystem, thus enhancing usage intention. Presently, the disjoint set of privacy-related protocols leads to confusion during implementation causing inconsistencies. IoT providers do not have a single universal privacy framework design for implementation in the IoT architecture and design. Industry wide legislations for such provisions will definitely help resolve such conflicts. It will also help assuage user privacy concerns regarding IoT-based consumer device adoption.

Reducing the gap between user wanted and manufacturer-designed features for IoT devices, can also improve trust in the ecosystem and thus improve adoption. Future research could explore design issues in ensuring these recommendations. Designing privacy-aware systems will improve perceptual control and hence information privacy. Empirical testing of the proposed model will help gain additional insights into this process of control traversal. Privacy concerns of industrial IoT devices do need to be studied too, to boost adoption of the whole connected IoT ecosystem.

References

[1]   Lewis D. Is Your TV Spying on You? [Internet]. *Forbes*. 2015 [cited 2016 Jan 25]. Available from: http://www.forbes.com/sites/davelewis/2015/02/10/is-your-tv-spying-on-you/

[2]   Steinberg J. These Devices May Be Spying on You (Even in Your Own Home) [Internet]. *Forbes*. 2014 [cited 2016 Jan 25]. Available from: http://www.forbes.com/sites/josephsteinberg/2014/01/27/these-devices-may-be-spying-on-you-even-in-your-own-home/

[3]   Ashton K. That "Internet of Things" Thing. *RFiD Journal*. 2009;22(7):97–114.

[4]   Tan CS. Samsung to Increase Consumer Electronics Chip Spending with the Expansion of IoT Ecosystem [Internet]. *IHS*. 2016 [cited 2016 Jan 25]. Available from: https://technology.ihs.com/570563/samsung-to-increase-consumer-electronics-chip-spending-with-the-expansion-of-iot-ecosystem

[5]   Gibbs S, Thielman S. Google Takes Aim at the Internet of Things with New Brillo Operating System [Internet]. *The Guardian*. 2015 [cited 2016 Jan 25]. Available from: http://www.theguardian.com/technology/2015/may/28/google-brillo-operating- system-internet-of-things

[6]   Westin AF. *Privacy and Freedom*. Atheneum, New York, NY, 1967.

[7]   Belanger F, Hiller JS. A Framework for e-Government: Privacy Implications. *Business Process Management Journal*. 2006;12(1):48–60.

[8]   Smith HJ, Milberg SJ, Burke SJ. Information Privacy: Measuring Individuals' Concerns about Organizational Practices. *MIS Quarterly*. 1996;20(2):167–96.

[9]   Fleisch E. What Is the Internet of Things? An Economic Perspective. *Economics, Management, and Financial Markets*. 2010;5(2):125–57.

[10]   Gershenfeld N, Krikorian R, Cohen D. The Internet of Things. *Scientific American*. 2004;291(4):76–81.

[11]   Roman R, Zhou J, Lopez J. On the Features and Challenges of Security and Privacy in Distributed Internet of Things. *Computer Networks*. 2013;57(10): 2266–79.

[12]   Weber RH. Internet of Things – New Security and Privacy Challenges. *Computer Law & Security Review*. 2010;26(1):23–30.

[13]   Sicari S, Rizzardi A, Grieco LA, Coen-Porisini A. Security, Privacy and Trust in Internet of Things: The Road Ahead. *Computer Networks*. 2015;76: 146–64.

[14]   EEC. "Commission recommendation of 12 May 2009 on the implementation of privacy and data protection principles in applications supported by radio-frequency identification.", available from (http://eur-lex. europa.eu/LexUriServ/LexUriServ.do?uri=OJ%3AL%3A2009%3A122% 3A0047%3A0051%3AEN%3APDF). Last accessed 15 June 2016 (2009).

[15]   Hoffman DL, Novak TP, Peralta M. Building Consumer Trust Online. *Communications of the ACM*. 1999;42(4):80–5.

[16]   Baronas A-MK, Louis MR. Restoring a Sense of Control during Implementation: How User Involvement Leads to System Acceptance. *MIS Quarterly*. 1988;12(1):111–24.

[17]   Veracode. The Internet of Things: Security Research Study [Internet]. *Veracode*. 2015 [cited 2016 Jan 25]. Available from: https://info.veracode. com/whitepaper-the-internet-of-things-poses-cybersecurity-risk.html

[18]   Carabelea C, Boissier O, Ramparany F. Benefits and Requirements of Using Multi-Agent Systems on Smart Devices. In: *Euro-Par 2003 Parallel Processing*. Springer, Berlin, 2003. p. 1091–8.

[19]   Tabak F, Smith WP. Privacy and Electronic Monitoring in the Workplace: A Model of Managerial Cognition and Relational Trust Development. *Employee Responsibilities and Rights Journal*. 2005;17(3):173–89.

[20]   Weick KE. *Sensemaking in Organizations (Foundations for Organizational Science)*. Sage Publications, Washington, DC, California, USA, 1995. p. 235.

[21]   Gioia DA, Chittipeddi K. Sensemaking and Sensegiving in Strategic Change Initiation. *Strategic Management Journal*. 1991;12(February):433–48.

[22]   Bartunek JM, Rosseau DM, Rudolph JW, DePalma JA. On the Receiving End: Sensemaking, Emotion, and Assessments of an Organizational Change Initiated by Others. *The Journal of Applied Behavioral Science*. 2006;42(2):182–206.

[23]   Lüscher LS, Lewis MW. Organizational Change and Managerial Sensemaking: Working through Paradox. *Academy of Management Journal*. 2008;51(2): 221–40.

[24]   Weick, Karl E., Goodman, Paul S. and Sproull, Lee S. "Technology as equivoque: Sensemaking in New Technologies". In: *Technology and Organizations*. The Jossey-Bass management series., (pp. 1–44). San Francisco, CA, US: Jossey-Bass, xxi, 281 pp. (1990).

[25]   Coburn CE. Collective Sensemaking about Reading: How Teachers Mediate Reading Policy in Their Professional Communities. *Educational Evaluation and Policy Analysis*. 2001;23(2):145–70.

[26]   Balogun J, Johnson G. Organizational Restructuring and Middle Manager Sensemaking. *Academy of Management Journal*. 2004;47(4):523–49.

[27]   Drazin R, Glynn MA, Kazanjian RK. Multilevel Theorizing about Creativity in Organizations: A Sensemaking Perspective. *Academy of Management Review*. 1999;24(2):286–307.

[28]   Alge BJ, Greenberg J, Brinsfield CT. An Identity-Based Model of Organizational Monitoring: Integrating Information Privacy and Organizational Justice. In: *Research in Personnel and Human Resources Management*, Emerald Insight, Vol. 25, 2006. p. 71–135.

[29]   Zuboff S. *In the Age of the Smart Machine: The Future of Work and Power*. Basic Books, USA, 1988.

[30]   Weick KE. Enacted Sensemaking in Crisis Situations. *Journal of Management Studies*. 1988;25(4):305–17.

[31]   DeSanctis G, Poole MS. Capturing the Complexity in Advanced Technology Use: Adaptive Structuration Theory. *Organization Science*. 1994;5(2):121–47.

[32]   Griffith TL. Technology Features as Triggers for Sensemaking. *Academy of Management Review*. 1999;24(3):472–88.

[33]   Macmillan IC, Mcgrath RG. Discover Your Products' Hidden Potential Discover Your Products' Hidden Potential. *Harvard Business Review*. 1996;74(3):58–73.

[34]   Davis FD, Bagozzi RP, Warshaw PR. User Acceptance of Computer Technology: A Comparison of Two Theoretical Models. *Management Science*. 1989;35(8):982–1003.

[35]   Venkatesh V. Creation of Favorable User Perceptions: Exploring the Role of Intrinsic Motivation. *MIS Quarterly*. 1999;23(2):239–60.

[36]   Venkatesh N, Davis FD. A Theoretical Extension of the Technology Acceptance Model: Four Longitudinal Field Studies. *Management Science*. 2000;46(2):186–204.

[37]   Venkatesh V, Morris MG, Davis FD, Davis GB. User Acceptance of Information Technology: Toward a Unified View. *MIS Quarterly*. 2003;27(3):425–78.

[38]   Milne GR, Bahl S. Are There Differences between Consumers' and Marketers' Privacy Expectations? A Segment- and Technology-Level Analysis. *Journal of Public Policy & Marketing*. 2010;29(1):138–49.

[39]   White TB. Consumer Disclosure and Disclosure Avoidance: A Motivational Framework. *Journal of Consumer Psychology*. 2004;14(1–2):41–51.

[40]   Harris L, Westin A. *E-commerce & Privacy: What Net Users Want*. Privacy & American Business, USA, 1998.

[41]   Awad NF, Krishnan MS. The Personalization Privacy Paradox: An Empirical Evaluation of Information Transparency and the Willingness to Be Profiled Online for Personalization. *MIS Quarterly*. 2006;30(1):13–28.

[42]   Venkatesh V. Determinants of Perceived Ease of Use: Integrating Control, Intrinsic Motivation, and Emotion into the Technology Acceptance Model. *Information Systems Research*. 2000;11(4):342–65.

[43]   Zuboff S. Big Other: Surveillance Capitalism and the Prospects of an Information Civilization. *Journal of Information Technology*. 2015;30(1):75–89.

[44]   Bandura A. Self-Efficacy Mechanism in Human Agency. *American Psychologist*. 1982;37(2):122–47.

[45]   Ajzen I. From Intentions to Actions: A Theory of Planned Behavior. In: *Action Control: From Cognition to Behavior*. Springer, Heidelberg, 1985. p. 11–39.

[46]   Ajzen I, Fishbein M. *Understanding Attitudes and Predicting Social Behavior*. Prentice Hall, Englewood Cliffs, NY, 1980. p. 278.

[47]   Fishbein M, Ajzen I. *Belief, Attitude, Intention and Behaviour: An Introduction to Theory and Research*. Addison Wesley, Reading, MA, 1975. p. 480.

[48]   Terry D, Terry DJ, Gallois C, McCarnish M. Self-Efficacy Expectancies and the theory of reasoned action. In: *The Theory of Reasoned Action Its Application to AIDS Preventive Behaviour*. Pergamon Press, UK, 1993. p. 135–52.

[49]   Phelps J, Nowak G, Ferrell E. Privacy Concerns and Consumer Willingness to Provide Personal Information. *Journal of Public Policy & Marketing*. 2000;19(1):27–41.

Chapter 4

SMARTIE: a secure platform for Smart Cities and IoT

José L. Hernández-Ramos, Dan García Carrillo,
Antonio Skarmeta, Fábio Gonçalves, Luis Cortesão,
Jens-Matthias Bohli, and Martin Bauer

Summary

The Internet of Things (IoT) is intended to transform our daily lives through the convergence of significant efforts from different disciplines. Under the vision of a hyper-connected world, security and privacy must be properly tackled by holistic approaches, since they are considered as key factors for a large-scale deployment of new IoT-enabled services and applications. Towards this end, the focus of the SMARTIE project is the application of security and privacy mechanisms to different use cases under the umbrella of a high-level architecture with a strong emphasis on security and privacy. The instantiation of this architecture is aimed to facilitate the creation of valuable and innovative services for citizens in the upcoming generation of Smart Cities, while security and privacy are preserved.

4.1 Introduction

The Internet of Things (IoT) [1] has become a widely used term to make reference to a global network of interconnected smart objects that promises to transform our surrounding environment. The vision of a hyper-connected world promises to overcome many socio-economic barriers in the coming years to transform technology into a ubiquitous element of our everyday lives. This view is being materialized by the convergence of technologies and multidisciplinary efforts, which are leading to the development of a smarter world bringing a significant benefit for people.

In this context, Smart Cities [2] represent emerging global ecosystems of interconnected smart objects, in which security and trust are essential for the success of innovative and valuable services to be leveraged by citizens. Data collected in a smart city platform will contain personal data that fall under data protection regulation, thus it needs to be protected by technical security means. Beyond protecting the privacy of citizens, there is a strong need to protect the information in order to reduce the risk of data theft that can lead to identify fraud and financial damage. Therefore, access to critical components needs to be protected in order to avoid disruption of the public infrastructure operation. In particular, actuators will allow hackers to extend the outreach of their attacks beyond Information Technology (IT) systems into the real world. The actual damages caused by possible threats can range from small interferences within the system to personal losses/exposure of private information. With the advent of Big Data era, the risk and impact of security or privacy threats will get a bigger impact for people.

Under these premises, the application of innovative and suitable approaches for security and privacy in smart cities is essential. Common network security tools, such as firewalls, monitoring systems or typical access control mechanisms, might not be enough to prevent sophisticated attacks due to the distributed nature of the IoT. It is essential that security is directly designed into the infrastructure rather than being added as an extra plug-in, by considering security and Privacy by Design (PbD) principles [4]. The challenge will be to design solutions where no single server has significant power to control the infrastructure or to access significant amounts of data. Thus, security solutions are required on all layers, from the sensor over the network, gateways up to the data storage and processing infrastructure. In this sense, the main goal of the EC-funded project SMARTIE (SMArter ciTIEs data management) [3] is to create a distributed framework for IoT-based applications storing, sharing and processing large volumes of heterogeneous information. This framework is envisioned to enable end-to-end security and trust in information delivery for decision-making purposes following data owner's privacy requirements. SMARTIE follows a data-centric paradigm, which will offer highly scalable and secure information for smart city applications. Its core features will be the information management and services plane as a unifying umbrella, which will operate on heterogeneous network devices and data sources, providing advanced secure information services to enable valuable higher-layer applications. SMARTIE is focused on the development of different technologies to fulfil the needs of the different users of such a platform with respect to security, privacy and trust. Specifically, several components have been already identified as a step forward to cope with the security and privacy requirements of a secure platform, they are being developed within the scope of the project. The core security components focus around encryption for data privacy, policy-based access control, and means for intrusion detection. However, more components providing further services required in the IoT in a secure way, e.g. discovery or platform management, are developed within the project.

In the following, we review related work, including a brief introduction to the Architecture Reference Model (ARM) of the IoT-Architecture project (IoT-A). In the main section, we provide a list of the main innovative components that are developed and used within SMARTIE. The list is structured according to functional groups (FGs) that are defined in the IoT-A ARM. This set of FGs can be considered common to IoT systems as it gathers the essential functionality of an IoT system. We then describe the scenarios where we plan to test the components and finally conclude.

4.2 Related work

The deployment of new IoT services requires to support aspects such as PbD and data minimization principles, in order to give people maximum control over their personal data. These requirements cannot be addressed at late stages during the development of IoT applications; they must be tackled during their design in order to build innovative applications for citizens by abstracting the underlying technologies, since IoT stakeholders will only accept services which are based on secure and privacy-preserving infrastructures. In recent years, the fragmentation of IoT technologies and protocols has led to the need for the definition of an abstract architecture in order to capture the inherent needs and requirements of the IoT ecosystem. Indeed, this issue has been widely considered as one of the main barriers for a large-scale adoption of this emerging paradigm. Several European efforts, such as the Internet of Things European Research Cluster (IERC),[1] have tried to fill this gap in order to define a harmonized vision that is agreed by academia and industry in order to build a more interoperable IoT. In fact, several European projects such as BUTLER,[2] SENSEI[3] or IoT6[4] represented multidisciplinary efforts to deal with different aspects of IoT. Specifically, BUTLER's main goal was to enable the development of secure and smart life assistant through the integration of different IoT devices and communication technologies, in order to provide a homogeneous access to the underlying heterogeneous networks. Moreover, IoT6 was intended to design an IPv6-based service-oriented architecture, in order to achieve a high degree of interoperability between different applications and communication technologies.

Under the umbrella of these initiatives, IoT-A [5] was a large-scale project focused on the design of an ARM to instantiate IoT architectures through a set of specific tools and guidelines. The main goal of this reference architecture was to optimize the interoperability among isolated IoT applications to create a global ecosystem of services under a common understanding. It was intended to promote initiatives adopting ARM as the baseline for other IoT architectures by enabling the reuse of components and interactions among them. While previous efforts were not focused on security and privacy aspects, the SocIoTal EU project[5] aims to provide a generic security framework addressing security and privacy requirements at different levels and fostering the participation of citizens to use valuable services with novel underlying security mechanisms. SocIoTal fosters the creation of a socially aware citizen-centric IoT by ensuring privacy and trust are embedded in a privacy-preserving context-sensitive framework for IoT devices which uses adequate security enablers. The resulting architecture represents a significant ongoing effort to build a more

---

[1] http://www.internet-of-things-research.eu/
[2] http://www.iot-butler.eu/
[3] http://www.sensei-project.eu/
[4] http://iot6.eu/
[5] http://sociotal.eu/

secure and privacy-preserving IoT. Also based on ARM, SMARTIE architecture has a strong emphasis on security and privacy aspects, through the instantiation of ARM FGs and its application to different use cases in the context of the next generation of Smart Cities.

4.3   IoT-A Architecture Reference Model

When facing a design involving the features and challenges of the IoT, working with a solid base is important. The IoT is often seen as vertical silos without interoperability, and using governance models that do not help with collaboration or interoperability. A good way to proceed is to use an established model that aids to identify the unique challenges that IoT systems bring. The IoT-A methodology is chosen for SMARTIE because it enables to concentrate the effort of the development on the main aspects of the project, while reusing concepts already established. IoT-Architecture (IoT-A; IP EU project from 2009 to 2012) [6] proposes a baseline design with an ARM, which aims to lower the aforementioned barriers such as interoperability or scalability that IoT designs have to face. This ARM uses basic building blocks together with guidelines and principles that are key in the design of protocols, interfaces and algorithms for an IoT system. This methodology and guidelines enable the design of an IoT system providing key features such as scalability and interoperability. Another important aspect of the IoT-A ARM is that it provides a language and concepts that improve the communication between the different partners. In addition, such a common language and semantics can be used for comparing the different approaches of each partner. All this is accomplished by means of a functional view and an information view, which, as we will explain later, consist of decomposition diagrams, interfaces, sequence charts, technical use cases and interaction examples. This methodology and guidelines will be summarized in the next section.

The IoT-A methodology for designing an IoT system uses a set of views (physical, context, functional, operational and deployment) that, in the end, define an architecture. The complete definition of the methodology can be found in the Deliverable D1.5 provided by the IoT-A project [7]. IoT-A provides a list of requirements that can be addressed by any IoT system, including functional and non-functional requirements. Based on this list of requirements, the IoT-A ARM process recommends, using it as a guideline, to derive specific requirements for the system. The elicited requirements demand some functionality that has to be provided by the system. This functionality can be grouped so that more or less complex components are generated. The groups of functionality can be related to aspects such as scalability, availability, security, performance, etc. In this way, it is easier to specify the components that may already exist or have to be designed, to address the requirements. The ARM Functional View (Figure 4.1) identifies FGs and FCs from the previous stage, the requirement process. This set of FGs can be considered common to IoT systems as it gathers the essential functionality of an IoT system.

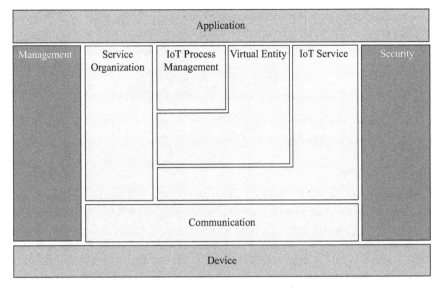

Figure 4.1 IoT-A Architecture Reference Model (ARM)

Specifically, IoT-A ARM presents nine FGs detailed in D1.5 of IoT-A [7, p. 108]. They are summarized as follows:

- Application: it represents the user or the application acting on behalf of the user.
- Device: it represents the target of the interaction, e.g. a sensor, source of information, a tag, actuator (modifier of the state of the system), etc.
- IoT Process Management: it identifies the new concepts or interfaces that are needed to gear the traditional processes towards the peculiarities of the IoT presents.
- Service Organization: it is in charge of managing and orchestrating services at the different levels of abstraction they are offered on.
- Management: it deals with the management of the IoT devices' configuration, monitoring, reporting, faults, etc., as well as the administration of the IoT system as a whole.
- Virtual Entity: it provides the functions necessary for interacting with the IoT system through Virtual Entities (VEs). In the same way, it provides the necessary functionalities for discovery and lookup of the services being offered for the aforementioned VEs.
- IoT Service: it gathers the IoT Services and their functionalities such as name resolution, discovery and lookup for those services.
- Communication: it serves as an abstraction that models the interaction schemes providing a common interface to the IoT Service Functional Group.
- Security: it is responsible for fulfilling security and privacy requirements regarding the IoT system.

The FGs are composed of different Functional Components (FCs). IoT-A describes the components' interfaces and functions, providing a means to design the interactions among the components; what is known as the information view. In the next section, we will follow the functional view of ARM to describe the components of SMARTIE.

4.4   SMARTIE architecture

4.4.1   Functional view

As described in the previous section regarding the IoT-A ARM, the functional view is used to describe a set of FGs and FCs that are derived from the requirements process and that can be found in several IoT systems. Figure 4.2 shows how the SMARTIE components are organized in the different FGs. Following the figure, we describe each SMARTIE component in relation with its FG and also the interaction with other components. The focus in this section's descriptions is on the reasoning of the placement in FGs and design decisions for the particular component.

4.4.2   Application Functional Group

Although the Application Functional Group is out of the scope of the IoT-A ARM, we present the components that offer high-level functionality to the IoT system. In the SMARTIE project, the Application Functional Group is composed of Intrusion Detection System (IDS)–Data Distribution and Smartdata Service Enablers for third-party Applications (SE/App).

4.4.2.1   IDS–Data Distribution

Attacks on IoT devices have to be reported to other devices and/or the network operator. The IDS is used to scan the network traffic for intrusions to the network and to report unknown or unwanted traffic to the network operator. Therefore, detected events are distributed to other devices. The data distribution parts of the IDS component immediately share new security-relevant information, mainly using the lightweight secure CoAP (lwsCoAP) component or tiny Data Storage Management (tinyDSM) distributed storage features.

4.4.2.2   Smartdata Service Enablers for third-party Applications

This component contains the logic and algorithms responsible for producing relevant context for the application from raw messages. These are mainly third-party applications, or services, running on top of the platform, that use, process, etc. the information collected by the platform. Services provided by components such as Privacy-Preserving Event Detection and Correlation, and Privacy-preserving geofencing (PrivLoc) may be incorporated in the platform as SEs, enabling enhanced functionalities through interaction with other available SEs.

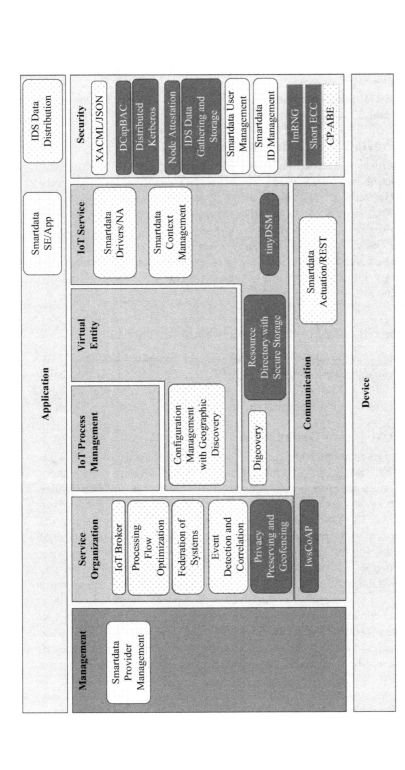

Figure 4.2   SMARTIE components for each functional group

4.4.3   Management Functional Group

The Management Functional Group offers features to the IoT System features to reduce costs, attend unforeseeable usage issues, adding flexibility to the overall system. In SMARTIE, the Management Functional Group contains the Smartdata Provider Management.

4.4.3.1   Smartdata Provider Management

The Smartdata Provider Management is a set of functionalities that gives other platform components, or third parties, the possibility to create, modify or delete providers. All the entities (i.e. sensors, applications, etc.) that generate information that should be collected by the platform are identified as Providers. To send information to the platform, a Provider has to be registered previously and authorized to send such data; otherwise, the information sent by the source is not taken into consideration. This component interacts with Directory Services – Resource Directory with Secure Storage and Digcovery. Functionalities such as semantic-based search may be provided by the component.

4.4.4   Service Organization Functional Group

This FG acts as a communication hub among several components of other FGs. As the basic communication concept in the IoT ARM, Service Organization composes services of different level of abstraction. In SMARTIE, it is composed of three components: Processing Flow Optimization, IoT Broker and Federation of Systems.

4.4.4.1   Processing Flow Optimization

The Processing Flow Optimization creates processing flows according to an optimization criterion, taking into account constraints. In a processing flow, sources (e.g. sensing services on sensor nodes) provide information that is processed by one or more information processing services on computing nodes. The information then flows from the sources through the information processing services, providing the required information to the requester. The selection of sources and the placement of information processing services on computing nodes are optimized according to an optimization criterion, e.g. required bandwidth or required processing power. The Processing Flow Optimization needs to be able to discover suitable sensor nodes with sensing services, processing service software components together with their requirements and compute nodes, on which the processing service software can be deployed. The Configuration Management with Geographic Discovery can provide the former. The latter has to be developed as part of the Processing Flow Optimization. This component will be integrated with the IoT Broker, so the information is on a VE level.

4.4.4.2   IoT Broker

The IoT Broker provides applications with one single access point for accessing IoT-related information from a possibly large set of IoT sources like sensors, but also information processors. The IoT Broker accesses the information from the different sources, aggregates it and returns it to the requesting application. Information sources have to provide their information using the Next Generation Services Interface (NGSI) Data Model for context information [8]. The IoT Broker supports synchronous one-time requests for accessing information, as well as asynchronous subscribe/notify interactions, where notification conditions specify when notifications (updates) are sent. The services it interacts with belong to the Virtual Entity Functional Group and it makes use of the Configuration Management with Geographic Discovery.

4.4.4.3 Federation of Systems

The Federation of Systems (FoS) allows temporary cooperation of independent systems in order to provide a given service, which cannot be provided by the involved subsystems separately. This component will be related to the Smart City Information Centre Use Case described in D2.1 [9]. It is an integrative use case showing the possibilities of cooperation of independent systems deployed in the Smart City in order to provide new functionalities. The FoS requires the tools looking for devices and services matching its requirements (e.g. through Digcovery, IoT Broker, Configuration Management and Resource Directory components). Also, access control needs to be ensured (e.g. through the interaction with Distributed Kerberos and Distributed Capability-Based Access Control (DCapBAC) components). Moreover, the communication within the FoS is intended to be secured through the use of Ciphertext-Policy Attribute-Based Encryption (CP-ABE) component and short Elliptic Curve Cryptography (shortECC) with the logistic map Random Number Generator (lmRNG). In addition, Node Attestation and Intrusion Detection components will be used to be able to detect the malicious or tampered devices and also the anomalies in the network traffic. Finally, in order to detect unusual measurements as soon as possible, Privacy-Preserving Event Detection and Correlation and tinyDSM components will be employed.

4.4.5 Virtual Entity Functional Group

The Virtual Entity Functional Group provides functionality for interacting with the IoT System based on VEs, as well as functionality such as discovery and lookup of services that provide information about VEs. Furthermore, it enables functionality for managing associations, finding dynamic associations and monitoring their validity. In SMARTIE, the Virtual Entity Functional Group contains the Configuration Management with Geographic Discovery.

4.4.5.1 Configuration Management with Geographic Discovery

The Configuration Management with Geographic Discovery is part of the Virtual Entity Functional Group, and more specifically it represents an instantiation of the VE Resolution Functional Component as described in the IoT Functional View presented in the IoT-A Deliverable D1.5 [7]. This component allows the lookup and discovery of information sources, e.g. sensing services that provide certain information, such as the indoor temperature of a room or the speed of a bus. For the discovery, a geographic scope can be specified using geographic coordinates, e.g. discover all services providing information about busses that are currently within a geographic area given by a geographic segment specified by the geographic coordinates of two opposite corners. Information sources register with the Configuration Management, specifying what information they can provide through the NGSI Data Model [8].

4.4.6 IoT Service Functional Group

The IoT Service Functional Group enables functionalities and services, such as discovery, name resolution and lookup of IoT services. In SMARTIE, the IoT Service Functional Group is composed of the following components: Digcovery, Resource Directory with Secure Storage, tinyDSM, Privacy-Preserving Correlation and Privacy-Preserving Geofencing, and also the Drivers/NA and Context Management of the Smartdata platform.

4.4.6.1 Digcovery

Digcovery is a framework that enables scalable and automated registration of devices for IoT environments using well-known protocols such as the Constrained Application Protocol (CoAP) [10] and CoAP Resource Directory. It provides the functionality of an information repository for the different devices in the system. Digcovery allows keeping the information about these devices updated and available to the users and the overall system. A common and previously agreed semantic has to be used in order to homogenize functionality and capabilities of the sensors. The use of CoAP as protocol for interacting with Digcovery enables the constrained devices to communicate with Digcovery and obtain information about services that they might need.

4.4.6.2 Resource Directory with secure storage and user authentication

The Resource Directory (RD) [11] component provides information on the available resources and services within the system that have been previously registered and enables efficient search and discovery in a secure manner. The resources are registered in the RD using specific message in eXtensible Markup Language (XML) format using Hypertext Transfer Protocol (HTTP) POST. This message contains metadata about the resource and/or service, capabilities, location and other relevant data. Subsequent search and discovery is performed using XML message sent using HTTP GET containing appropriate filters, indicating desired services or resources. It relies on the secure storage and other SMARTIE components, in order to enable secure access to the dynamic information about all available resources in the system at a given time only to authorized users and components.

4.4.6.3 Tiny Data Storage Management

The tinyDSM middleware provides a shared data storage for the nodes in the wireless sensor network. The data pieces are defined as variables, each with a specified type and policy controlling the behaviour of the middleware regarding handling of the variable. The sharing of the variables is based on replication and the policies specify its details in terms of the replication area and consistency parameters. The replication helps to assure the data availability in case of node failure and provides a consistent view of the replicas. The configuration is application specific, static and common for all the nodes in the network. Reliable shared data storage with data monitoring makes it possible for the nodes to cooperate more autonomic and independent from the central station by injecting more intelligence into the network.

4.4.6.4 Privacy-preserving geofencing

PrivLoc offers secure location-based services, in particular a secure geofencing service. It uses a standard spatial database for geolocations (e.g. postgreSQL, MySQL, mongoDB) without modifications. The core of PrivLoc is a lightweight encryption procedure for location data that preserves operations like intersecting geometric objects and nearest neighbour search within proximity. In a typical geofencing setting users can subscribe to an area in order to get notification if any of the tracked objects enter or leave the area. A spatial database hosted on the cloud receives regular movement vectors of the objects, and checks if these movements cross a geofenced area which has been subscribed to within the database.

4.4.6.5 Privacy-preserving event detection and correlation

This component allows privacy-preserving processing of sensor data. The success of this concept will not materialize just with a mere gathering of the information generated by the multiple sensors deployed everywhere in a smart city. An intelligent and powerful correlation among an enormous amount of data is definitely a must to fulfil such a goal. By linking, grouping, associating and/or clustering all the collected data given certain criteria, more meaningful and useful information, such as behavioural patterns, for instance, could be extracted. In order to preserve the citizens' privacy in the context of smart cities, we suggest to encrypt the data before being collected, in such a special way that its original content is never revealed to the information-gathering entity, but it is yet usable for certain purposes. In particular, it should be possible to perform a number of operations devoted to achieve a smart and useful correlation of such encrypted data.

4.4.6.6 Smartdata: Drivers/NA

A driver is considered as a component able to translate the information (e.g. the payload of the information sent by a CoAP sensor) to a well-known language understandable and manageable by the platform. Drivers are components that can dynamically be added to the platform by third-party entities. There are components that run inside the platform, but are not part of the platform base components. This component works at a low-level layer, providing the translation mechanisms between the platform and devices.

4.4.6.7 Smartdata: Context Management

The Context (Ctx) Management component deals with the routing mechanism necessary to forward the messages from the module that receives it to the module that should consume the message. It provides the queuing and routing functionalities.

Ctx Management uses the components in the Security Functional Group to guarantee that the providers are authenticated and authorized to publish data to the platform.

4.4.7 Security Functional Group

The Security Functional Group is in charge of enabling privacy and security of the IoT system. In SMARTIE, this is the most important FG that contains many relevant components, such as DCapBAC, eXtensible Access Control Markup Language (XACML)/JavaScript Object Notation (JSON), CP-ABE, Sticky Policies Access Control and Encryption (SPACE), IDS Data Gathering and Storage, Node Attestation (IMASC), Distributed Kerberos, shortECC, lmRNG and the Smartdata Components User Management and ID Management.

4.4.7.1 Distributed Capability-Based Access Control

The DCapBAC [12] component is based on the use of authorization tokens, containing access privileges that were previously granted to the holder, as well as a set of access conditions to be locally verified at the end device. Under the main foundations of AuthoriZation-Based Access Control (ZBAC) [14], DCapBAC allows a distributed approach in which IoT devices are enabled with authorization logic by adapting the communication technologies and representation format. Specifically, it makes use of the JSON [15] as representation format for the capability token, which are attached on access requests by using CoAP [10]. Furthermore, it defines an optimized ECC library, in order to provide an access control mechanism for the IoT, while end-to-end security is preserved.

4.4.7.2 XACML/JSON

XACML [16] is a framework consistent with Attribute-Based Access Control (ABAC) [18] that uses policies to express rich and fine-grained access control decisions. It is based on the XACML [17] by giving the possibility to define access control policies in JSON, instead of XML. Furthermore, access control policies have to be managed in an efficient way to handle all rules and policies, having into account possible conflicts, duplication, etc. among them. XACML provides access control functionality following the current state-of-the-art standard in this area. Moreover, the inherent features of IoT and the amount of requests that are expected of IoT Systems require efficiency to be taken into account. This is why the use of JSON in conjunction with XACML is considered as part of the functionality of this component.

4.4.7.3 Ciphertext-Policy Attribute-Based Encryption

The CP-ABE [13] is a cryptographic scheme that links access control and encryption. The main foundation of CP-ABE is an algorithm to encrypt information under a certain policy of identity attributes. Secret keys are associated with sets of attributes and given to the users how are supposed to have the respective attribute. This enables dissemination of information using different information sharing models where the producer of the information is always in control of how the information can be accessed. Furthermore, CP-ABE can be used in scenarios where multiple receivers are involved. With CP-ABE, the information does not need to be encrypted individually for each receiver, but only once according to an access policy. The scheme itself ensures that all authorized recipients, holding secret keys with the right attributes, can decrypt the message. Furthermore, it could be integrated with ABAC through the use of the XACML/JSON component in order to enhance the consistency of the security functionality.

4.4.7.4 Sticky Policies Access Control and Encryption

As an alternative mechanism to CP-ABE, the SPACE component will allow IoT devices to store the information in an untrusted environment. This component is intended to provide a scalable and granular information access control mechanism, allowing end-to-end information encryption, information access to specific elements within a group, as well as secure information storage. SPACE is built on top of the Sticky Policies concept [19], which is a policy-based encryption scheme that allows an entity to encrypt data according to a policy file, in such a way that only who meets the policy file requirements has access to the information. It also enables an entity to store data in an untrusted database in a secure way. Specifically, the policies are defined in XACML 3.0, which allows the expression of very complex rules and various types of access control, such as ABAC, and easy integration with the XACML/JSON component, in order to reduce the size of policy files.

4.4.7.5 Short Elliptic Curve Cryptography

The shortECC library provides the encryption and digital signature functionalities for low-power devices, e.g. wireless sensor nodes. The approach uses elliptic curves with a key length between 32 and 64 bits, but can also use standard ECC key lengths depending on the trade-off between the security level and performance. It can be used by another FCs requiring the mentioned cryptographic functionalities. Specifically, it can be employed by components requiring secured communication with resource-constrained devices and integrated with CoAP in order to provide basic security properties. Furthermore, this component makes use of tinyDSM for key management and lmRNG in order to generate the pseudorandom numbers.

4.4.7.6 Logistic map Random Number Generator

The lmRNG allows the generation of cryptographic secure pseudorandom numbers on low-constrained devices, e.g. wireless sensor nodes. Furthermore, for this component, it is envisioned the use of a hardware module for the seed generation. Its outputs need to fulfil the National Institute of Standards and Technology (NIST) [20] requirements according to entropy sources. The modular arithmetic performed by the lmRNG needs to be adjusted to the finite field order in which the cryptographic operations are performed. It can be used by security libraries, e.g. shortECC or by another FCs requiring pseudorandom numbers for security aspects.

4.4.7.7 Node Attestation (IMASC)

IMASC is an integrity measurement framework that makes use of a cheap and common secure hardware, the SmartCard, to provide a trusted running environment for most of the embedded devices that has suitable port support. Since the SmartCard is a secure microcontroller that is very difficult for the attacker to hack, we use it as a trust anchor in the architecture of the device node to maintain the integrity of its software stack. The firmware of the device will be tailored to a measure-before launch execution scheme, so that tampered code or unknown libraries will be detected and audited before the device decided to execute it. Meanwhile, the SmartCard protects the integrity of the measurement results so that a Trusted Third Party can attest each node remotely.

4.4.7.8   Distributed Kerberos

Distributed Kerberos is a component to compute Kerberos-like authentication tokens. The service can be enriched with authorization, e.g. with capabilities based on the DCapBAC component to obtain lightweight capability tokens for authentication and authorization. With symmetric cryptography, the choice of suitable authentication and authorization schemes is limited. A popular choice is the Kerberos protocol, which is, however, designed for an enterprise scenario, where all resources and the central Kerberos servers are in the same domain. Distributed Kerberos increases the trust in the Kerberos server by distributing it over two entities, e.g. two servers or even two organizations. As long as at most one server part is compromised, all obtained Kerberos tokens follow the access control policy. For a successful attack, both servers would need to be compromised. Therefore, Distributed Kerberos can be used as a publicly available service across domains.

4.4.7.9   IDS data gathering and storage

IoT devices will likely be exposed to attacks from the Internet. Encryption and authentication mechanisms may not be sufficient to protect those kinds of distributed low-power devices. The IDS component detects on-going attacks on IoT devices. Therefore, the data gathering and storage parts of the IDS component scan the network traffic (mainly from lwsCoAP component) for intrusions to the network/devices and may correlate possible events from other event sources (applications/tasks) for security-relevant data. It gathers and stores part of the security-relevant data to build a knowledge base for further detection. This component gathers and stores data to build a knowledge base for detection. Data storage may include local storage possibilities as well as distributed tinyDSM storage features.

4.4.8   Communication Functional Group

The Communication Functional Group is an abstraction used for modelling the variety of scheme interactions derived from several technologies belonging to IoT Systems. This FG also provides a common interface to the IoT Service Functional Group. In SMARTIE, it is composed of lwsCoAP and the Actuation/REpresentational State Transfer (REST) component of the Smartdata platform.

4.4.8.1   Lightweight secure CoAP

Existing secure CoAP mandates the use of Datagram Transport Layer Security (DTLS) [21] as the underlying security protocol. However, DTLS was originally designed for comparably powerful devices that are interconnected via reliable, high-bandwidth links, which is often not the case. The challenge is to deploy simplified encryption methods as part of the CoAP, which can be executed on very constrained devices. The lwsCoAP component is used to provide secure data channel between the IoT devices and the back-end cloud platform by using secure channel and employing lightweight encryption schemes. The core of the security system is the cryptographic primitive based on ECC, which can be successfully scaled up and down to provide variable level of protection at the expense of using more or less resources (i.e. processing power, memory, generated overhead). The solution is based on ISO/IEC 29192 standards, which aim to provide lightweight cryptography for constrained devices, including, block and stream ciphers and asymmetric mechanisms. This method is further optimized in order to reduce the key size and makes the algorithm more efficient.

4.4.8.2 Smartdata Actuation/REST

This component is responsible for the communication between the platform and the device. It translates the abstract message known by the platform to the specific message that it is understandable by the corresponding device. This component is able to route the message from the platform (or from the SEs) to the actuators. Specifically, it is intended to translate the messages, from an abstract view to an actuator's understandable message.

4.4.9 Architecture configurations

The SMARTIE components described in the previous section have the same or at least similar functionalities, but differ in their characteristics, which make them suitable for different scenarios. In this section, for a few high-level scenarios that have specific characteristic and requirements, we identify a set of components especially suited for the given setting. The scenarios only serve as examples; more complex scenarios may have multiple heterogeneous requirements, which may result in having different levels and a set of components for each level. In this case, integration between the different levels is needed.

4.4.9.1 Constrained devices scenario

This scenario is characterized by a number of constrained devices, e.g. sensor nodes, and a limited not (heavily) constrained infrastructure, e.g. a single gateway. The constrained devices interact with the gateway, but also with each other to achieve the desired functionality. An example for such scenario could be environmental monitoring in a remote area with limited infrastructure. The devices are constrained with respect to computational power, energy and communication bandwidth. For this reason, there are few suitable off-the-shelf components available that could be used, especially if specific security requirements are to be fulfilled. Figure 4.3 shows a number of high-level functionalities required in this scenario. For each of these, the selected SMARTIE components and libraries are provided, together with an explanation why they have been selected.

4.4.9.2 Non-constrained IoT services

In this case, the scenario is characterized by IoT services based on devices with sensors and actuators. Devices are not (heavily) constrained, e.g. involving smart

High-level functionality	Components and libraries	Explanation
Communication	IwsCOAP, ImRNG, shortECC, Smartdata Actuation/REST	Constrained devices have to securely communicate with existing infrastructure and among themselves, off-the-shelf secure communication is too expensive for constrained devices.
Discovery	Digcovery, IwsCOAP	Sensors need to register themselves and find out about services, so the interaction with the discovery component must be light-weight.
Storage	tinyDSM	Constrained devices need very light-weight storage
Device Management	Smartdata Provider Management	Enables to create, modify or delete providers. All devices, constrained or not constrainer, must be registered in order to send data to the platform.
Access Control	DCapBAC, XACML/ JSON, Smartdata Id management, Smartdata User management	Constrained devices require light-weight and distributed mechanism to control the access of their services and resources. Smartdata Id management incorporates components to provide Access control to constrained and non-constrained devices.
Authentication	Distributed Kerberos, Smartdata Id management, Smartdata User management	Kerberos tokens do not require public-key cryptography and can be processed on restricted nodes. All entities must be authenticated in order to access the platform. Smartdata Id management incorporates components to provide Access control to constrained and non-constrained devices.
Driver	Smartdata Drivers/NA	As many devices and protocols can be used in order to communicate with the platform, a driver is needed to translate the information to a well-known language in order to guarantee its persistence.
Routing	Smartdata Ctx Management	A routing mechanism that moves the messages from the module that receives it to the module that consumes the message.
Information Processing	Smartdata SE/App	Processing the information from the application raw messages in order to provide relevant context for the applications.

Figure 4.3   Components for constrained devices scenario

phones, and there is some infrastructure. There may be higher security requirements, e.g. a requirement for device attestation. An example for such scenario could be a medical home care application, where a smaller number of sensors and actuators are deployed and it is important that information is secured and tampering with devices does not go undetected. Figure 4.4 shows a number of high-level functionalities required in this scenario. For each of these, the selected SMARTIE components and libraries are provided, together with an explanation why they have been selected.

High-level functionality	Components and libraries	Explanation
Communication	Off-the-shelf secure communication, e.g. CoAP	Envisioned devices can handle suitable cryptographic primitives and protocols.
Discovery	Resource Directory	Resource-level registrations suitable for IoT Services, a secure database as provided by the Resource Directory requires some infrastructure.
Device Attestation	IMASC	Non-constrained devices can support device attestation.
Device Management	Smartdata Provider Management	Enables to create, modify or delete providers. All devices, constrained or not constrainer, must be registered in order to send data to the platform.
Access Control	DCapBAC, XACML/ JSON, Smartdata Id management, Smartdata user management	Non-constrained devices can support distributed access control. Smartdata Id management incorporates components to provide Access control to constrained and non-constrained devices.
Authentication	Smartdata Id management, Smartdata user management	All entities must be authenticated in order to access the platform. Smartdata Id management incorporates components to provide authentication to constrained and non-constrained devices.
Driver	Driver/NA	As many devices and protocols can be used in order to communicate with the platform, a driver is needed to translate the information to a well-known language in order to guarantee its persistence.
Routing	Smartdata Ctx Management	A routing mechanism that moves the messages from the module that receives it to the module that consumes the message.
Information Processing	Smartdata SE/App	Processing the information from the application raw messages in order to provide relevant context for the applications.
Location Services	PrivLoc	If the device has a GPS or similar sensor for locating itself, it can encrypt the location using PrivLoc for processing the data in the cloud.

Figure 4.4   Components for non-constrained IoT Services scenario

4.4.9.3   Large-scale IoT Deployment with VE level services

This scenario is characterized as a large-scale IoT scenario with a strong infrastructure. A higher abstraction level is supported that allows discovering and querying information on a VE level, e.g. request information like the occupancy of a room or the power consumption of a building instead of requesting the measured value of sensor A. Processing of information is required as aggregated results are requested and there are different players involved. Processing can be done on different nodes, including sensor nodes, gateways, special servers or the cloud, but there may be

High-level Functionality	Components and Libraries	Explanation
Communication	Off-the-shelf	No constraints, so existing secure communication can be used.
Discovery	IoT-Broker, Configuration Management	Configuration Management supports registration, look-up and discovery of VE-level services with geographic scopes; IoT Broker provides point of access for applications accessing information.
Processing	Processing Flow Optimization	Optimization of where processing takes place and how information flows through system, takes into account security constraints, e.g. trust level, availability of cryptographic keys.
Authentication	Smartdata Id management	All entities must be authenticated in order to access the platform. Smartdata Id management incorporates components to provide authentication for virtual entities.
Event detection	Event detection and data correlation	Event-detection or data correlation on encrypted data must be set-up on large processing nodes, e.g. in a cloud infrastructure. Sensors or their virtual representation will encrypt the data.

Figure 4.5   Components for large-scale IoT Deployment with VE level services scenario

different trust levels, which has to be taken into account for selecting appropriate nodes for processing. An example for such a scenario could be a smart city scenario with the city itself, an energy provider, a public transport company, etc. Figure 4.5 shows a number of high-level functionalities required in this scenario. For each of these, the selected SMARTIE components and libraries are provided, together with an explanation why they have been selected.

4.5   SMARTIE scenarios for IoT-enabled smart cities

There are several scenarios that in SMARTIE are used to represent typical use cases in which the use of IoT things would improve the current use of assets, resources and the performance of sensitive tasks.

4.5.1   Smart energy management

The goal of this use case is to provide a reference system able to manage intelligently the energy use in buildings, allowing a full control on the access to information and communications security. Energy efficiency in buildings requires the interaction between a number of actors and entities providing energy monitoring and consumption feedback, using automation systems, sensors and actuators, and carrying out economic strategies to save energy. In order to cover such requirements at city level, it is necessary to provide a common platform which lets users to be informed about energy use aspects, as well as the interaction with the system to define specific strategies for saving energy or control their own devices that are integrated in the platform. In this scenario, SMARTIE security components are responsible for controlling the access to the devices and information related to energy management. In particular, this is built on top of an identity management scheme, in order to control which entities can access what information and under what circumstances. The integration of these mechanisms with other intrusion detection techniques will help to detect misuse or abuse of the information that is required for this use case.

The pilot will be set in the Region of Murcia (Spain), where different city facilities will be monitored and managed by the SMARTIE platform to deal with energy efficiency at city level. Murcia already has several target facilities that can be considered as relevant to deal with energy consumption aspects, due to their contribution to the energy consumption at city level. Among others, public facilities such as schools, hospitals and public buildings are being monitored in term of their energy usage behaviour. Therefore, energy consumption levels associated to different city subsystems can be provided to become citizen and government aware about this aspect. Besides, strategies to save energy in such facilities can be defined, as well as future plans to improve energy efficiency in cities.

4.5.2 Public transport scenario

The system proposed in this Use Case aims to improve the management of the public transportation network in the city of Novi Sad (Serbia) starting from the Public City Bus Transport Network. The resulting scenario is intended to promote and encourage use of sustainable transport modes, saving costs and time to travellers. The pilot to be set will be based on enabling smart transport options for users of a public transport focusing initially on two routes within a city public bus transport network operated by a local transport company JGSP. Bus stops covering the two routes will be equipped with Augmented Reality (AR) markers in the form of an image (e.g. logo or QR code). Furthermore, fleet management devices will be placed on the appropriate buses, in order to track their location in real time. Users (travellers) will be able to use their smart phones, a dedicated application, and the AR marker at the bus stop, to find out the bus arrival time, as well as to request information about the best route to a certain destination depending on their criteria.

In this scenario, data generated by the fleet management devices are owned by the public transportation company, and the access to this data should be restricted to only authorized users. Furthermore, citizens will be generating private data to indicate the Global Positioning System (GPS) location, as well as their travel plans. This data will be stored within the cloud infrastructure and they should be also treated as sensitive information. Therefore, the access should not be made publicly available. Moreover, it should be prevented that any unauthorized fleet management device is connected to the system. Consequently, it is necessary to establish access control policies for citizens and fleet management connecting to the back-end cloud platform. Any communication within the system must be made secure using the appropriate methods, by taking into the account different layers within the system's architectural stack. In particular, security mechanisms should be able to address this issue whether operating on the powerful cloud back-end infrastructure, less powerful mobile phone platforms or resource-constrained devices. The system infrastructure will address and prevent any potential threat at different levels of the system by using the SMARTIE platform and solutions.

4.5.3   Traffic management scenario

One of the main future challenges for urban administrations will be the management of the constantly increasing amount of urban traffic, especially in the metropolitan areas. In nearly every larger town in the world, congestion situations affecting large areas of the inner city are a daily occurrence. This is not even problematic because of the higher amount of noise and pollution caused by the traffic. It also causes higher transport costs to the communities and decreases traffic safety considerably. The aim of this use case is to use the SMARTIE platform and solutions to improve the traffic situation, information level of the road user and traffic safety. Therefore, the existing traffic infrastructure in a region must be combined with the SMARTIE platform.

This scenario will enable the traffic management authorities to join different traffic data sources and actuators to improve traffic flow and traffic safety in the relevant area. Parts of this use case will flow into a pilot system in the city of Frankfurt Oder (Germany). In this case, the application of SMARTIE security foundations is crucial to ensure traffic data is coming from legitimate devices. Intrusion detection mechanisms and sophisticated encryption schemes (e.g. CP-ABE or SPACE) could be applied to control how location information is disseminated. The pilot will show the possibilities of the SMARTIE platform with a special attention to emergency situations. Therefore, the existing traffic infrastructure of Frankfurt will be especially considered in this use case description. The Traffic Green system is in operation in the city for over 10 years now. So the most important goal of this use case is to connect different independent systems, like the ones mentioned here via the SMARTIE platform to avoid abnormal traffic situations or dissolve them quickest possible if they have emerged.

4.5.4   City information centre scenario

The idea of a smart city information centre is to gather all the information about a city and provide access to it on a suitable abstraction level. The purpose is to give the responsible people a quick overview of what is going on in the city and whether there are issues that require attention, possibly indicating a level of urgency. To achieve this, suitable visualizations of the information are needed with the possibility to increase the granularity, i.e. to zoom into a problem area. The thematic dimensions that could be included are traffic, public transport and energy production consumption as covered by the SMARTIE use cases, and others. To understand the situation and how it develops, single data points showing the current situation are not sufficient, but the data needs to cover the spatial and temporal dimensions.

The smart city information centre can serve as the basis for all departments that need to keep an overview of what is going on in the city. As the result of privatization, this may also include companies that have taken over certain public services. The overall information provided goes beyond what would be available through the information sources of a single city department or company, e.g. for the traffic case additional information related to the cause of the problem like movement of crowds, flooding or a fire, could be highly relevant. In this case, SMARTIE security components will be in charge of the security and privacy-preserving mechanisms for data dissemination through the information centre. It will be based on different techniques for encryption (i.e. CP-ABE and SPACE), as well as the integration with authentication components (i.e. Distributed Kerberos) and DCapBAC (together XACML) for authorization. Finally, the city information centre can be used as a basis for coordinating activities of different agencies, e.g. the fire brigade and traffic management in the case of fires or floods. Having different thematic aspects accessible in the same place allows the automatic correlation of different information and creating alerts when a critical situation is detected. This could be especially relevant for an area like public safety.

4.6   Discussion

The idea of the IoT brings new challenges regarding security, privacy and trust. In the envisioned IoT-enabled Smart Cities, the potential challenges for a secure and privacy-preserving ecosystem need to be addressed by innovative mechanisms, in order to encourage the adoption of IoT services by citizens. The realization of this view requires multidisciplinary approaches that are able to cope with the needs of citizens, as well as the inherent features of IoT devices and platforms composing such scenarios. In this sense, the components of the EC-funded SMARTIE project provide a set of innovations and enhancements to address such challenges through their deployment on different real-world scenarios. SMARTIE architecture is based on the ARM from IoT-A, by instantiating the different FGs through technologies and protocols that are specifically intended to be deployed on IoT scenarios. Specifically, SMARTIE defines a set of security and privacy technologies that are embedded into the Security Functional Group. Such procedures are being implemented and deployed in different cities under the umbrella of different use cases, which are intended to demonstrate the feasibility of them. The resulting instantiated architecture from ARM represents a significant step forward to foster the development of secure and privacy-aware services in the context of the future Smarter Cities.

Acknowledgements

This work has been sponsored by European Commission through the FP7-SMARTIE-609062 EU Project.

References

[1]   Atzori, L., Iera, A., Morabito, G. "The Internet of Things: A survey". *Computer Networks*, 54(15), 2787–2805, 2010.

[2]   Caragliu, A., Del Bo, C., Nijkamp, P. "Smart Cities in Europe". *Journal of Urban Technology*, 18(2), 65–82, 2011.

[3]   Bohli, J. M., Skarmeta, A., Victoria Moreno, M., Garcia, D., Langendorfer, P. "SMARTIE project: Secure IoT Data Management for Smart Cities". In *International Conference on Recent Advances in Internet of Things (RIoT)*, Singapore, IEEE, 2015.

[4]   Langheinrich, M. "Privacy by design – Principles of privacy-aware ubiquitous systems". In *Ubiquitous Computing (Ubicomp)* (pp. 273–291), Atlanta, Georgia, USA, ACM, 2001.

[5]   Bassi, A., Bauer, M., Fiedler, M., *et al. Enabling Things to Talk. Designing IoT Solutions with the IoT Architectural Reference Model*, Springer, Berlin, 2013.

[6]   FP7 EU IoT-A Project, "Internet of Things architecture". Available from: http://www.iot-a.eu/, 2011 [Accessed 15 June 2016].

[7]   IoT-A Deliverable D1.5, *Final Architectural Reference Model for the IoT*, 2013.

[8]   Open Mobile Alliance, *NGSI Context Management Specification, Open Mobile Alliance*, 2015.

[9]   Shelby, Z., Hartke, K. SMARTIE Deliverable D2.1, *Use Cases*, 2015.

[10] Shelby, Z., Hartke, K., Bormann, C., *RFC 7252: The Constrained Application Protocol (CoAP)*, IETF (Internet Engineering Task Force), 2014.

[11] Shelby, Z., Koster, M., Bormann, C., van der Stok, P., *Internet-Draft: CoRE Resource Directory (draft-ietf-core-resource-directory-04)*, IETF (Internet Engineering Task Force), 2014.

[12] Hernández-Ramos, J. L., Jara, A. J., Marín, L., and Skarmeta, A. F., "DCapBAC: Embedding authorization logic into Smart Things through ECC optimizations". *International Journal of Computer Mathematics*, 93(2): 345–366, February 2016.

[13] Bethencourt, J., Sahai, A., Waters, B. "Ciphertext-policy attribute-based encryption". In *IEEE Symposium on Security and Privacy. SP'07* (pp. 321–334), Oakland, California, USA, IEEE, 2007.

[14] Karp, A. H., Haury, H., Davis, M. H. "From ABAC to ZBAC: The evolution of access control models". In *Proceedings of the Fifth International Conference on Information Warfare and Security* (pp. 202–211), Ohio, USA, 2010.

[15] Crockford, D. *RFC 4627: The Application/JSON Media Type for JavaScript Object Notation (JSON)*, IETF (Internet Engineering Task Force), 2006.

[16] Brossard, D. OASIS Draft, *JSON Profile of XACML 3.0 Version 1.0*, OASIS (Organization for the Advancement of Structured Information Standards), 2014.

[17] Rissanen, E. OASIS Standard, *eXtensible Access Control Markup Language (XACML) Version 3.0*, OASIS (Organization for the Advancement of Structured Information Standards), 2014.

[18] Yuan, E., Tong, J. "Attributed based access control (ABAC) for web services". In *IEEE International Conference on Web Services (ICWS)*, Orlando, Florida, USA, IEEE, 2005.

[19] Mont, M. C., Pearson, S., Bramhall, P. "Towards accountable management of identity and privacy: Sticky policies and enforceable tracing services". In *14th IEEE International Workshop on Database and Expert Systems Applications* (pp. 377–382), Prague, Czech Republic, Springer, 2003.

[20] Barker, E., Kelsey, J. *Recommendation for the Entropy Sources Used for Random Bit Generation*, Draft. NIST Special Publication, USA, 2012.

[21] Rescorla, E., Modadugu, N. *RFC 6347: Datagram Transport Layer Security Version 1.2*, IETF (Internet Engineering Task Force), 2012.

Chapter 5

Model-based security engineering for the Internet of Things

*Ricardo Neisse, Gary Steri, Igor Nai Fovino,
Gianmarco Baldini, and Lodewijk van Hoesel*

Summary

We propose in this chapter a Model-based Security Toolkit (SecKit) and methodology to address the control and protection of user data in the deployment of the Internet of Things (IoT). This toolkit takes a more general approach for security engineering including risk analysis, establishment of aspect-specific trust relationships, and enforceable security policies. We describe the integrated metamodels used in the toolkit and the accompanying security engineering methodology for IoT systems using these metamodels. We validate our approach through a case study of a real-world supply chain scenario where sensors are used to monitor the temperature and control environmental conditions of the transported goods. The toolkit is applied in the design of this case study, analysis of risks, and specification of security policy rules following the steps of our methodology. Finally, we also show how the specified security policies are enforced using technology-specific policy enforcement points.

5.1 Introduction

The control and protection of user data is a very important aspect in the design and deployment of the Internet of Things (IoT). This aspect is highlighted by the European Union data protection legislation, which states that privacy is a fundamental right that should be addressed. The heterogeneity of the IoT technologies, the number of the participating devices and systems, and the different types of users and roles create important challenges in the IoT context. In particular, requirements of scalability, interoperability, security, and privacy are difficult to address even with the considerable amount of existing work both in the research and standardization community.

In this chapter, we describe a Model-based Security Toolkit, named SecKit, which is integrated in a management framework for IoT devices [18, 20]. The main characteristic of the SecKit is a more general enterprise architecture approach for security engineering in IoT systems considering risk analysis, trust management, and security policy management. After the specification of the system models, a risk analysis using threat scenarios is performed, which leads to the design of countermeasures using security policy rule templates or establishment of trust relationships needed to mitigate the risks. From a policy management perspective, the SecKit supports specification, evaluation, and enforcement of security policies using an expressive policy language, efficient policy evaluation algorithm, and a distributed policy enforcement architecture [19]. Our main contribution is a description of the integrated SecKit metamodels, focusing on a security engineering methodology for IoT systems using these metamodels.

The SecKit has been developed in the context of the iCore Project [31], which has the goal of addressing the complexity and heterogeneity of different objects and technologies while maximizing the exploitation and provision of IoT objects and their services. The iCore Project proposes the abstraction of IoT using the Service concept, which is further refined in Composite Virtual Objects (CVOs) and Virtual Objects (VOs). A VO is a virtual representation of any Real-World Object (RWO) or digital object. A car, for example, can be represented as a CVO consisting of an engine, various sensors, and a communication system, which are all represented by VOs. Finally, a complete system or device can provide access to their capabilities, which are represented by a Service. This framework is related to concepts already proposed by other paradigms like the Service-Oriented Architecture and Object-Oriented databases.

We validate the SecKit metamodels, components, and methodology through a case study of a real-world supply chain scenario. The main objective of this scenario is to observe products in a supply chain by means of sensors (either attached to the products or statically fitted in warehouses or transport vehicles), detect anomalies (e.g., temperature) and provide feedback on the product conditions, or even adapt the environmental conditions (e.g., adjust temperature). Privacy and trust mechanisms are needed to allow/prevent data sharing across companies/stakeholders including support of different levels of access to data. For example, the product owner must be able to access information of his own products, but should not be allowed to access to information about other items in a warehouse or transport vehicles.

The enforcement of security policy rules is done using the SecKit runtime components, which are applied at the VO level to prevent the monitored information from being accessed by unauthorized entities. The SecKit intercepts transparently the messages exchanged with VOs at technology-specific implementation layer (Message Queue Telemetry Transport – MQTT) and enforces the security policies specified by the iCore system operator or by users. From a security enforcement perspective, we use a Policy Enforcement Point (PEP) deployed in the MQTT broker, which is a widely adopted technology to enable the communication between IoT devices by the academia and industry. The PEP sends event notifications to the Policy Decision Point (PDP), which is the component responsible for the evaluation of the security policy rules.

The remaining of this chapter is structured as follows. Section 5.2 describes the related work. Section 5.3 presents the iCore IoT framework. Section 5.4 describes the SecKit methodology steps and the elements defined in each of the steps. Section 5.5 illustrates the smart business case study, which is used in Section 5.6 to present the

SecKit methodology for security engineering. Section 5.7 concludes this chapter with a summary of our approach, a discussion on our results, and future directions.

5.2 Related work

After having analyzed existing approaches for security and privacy in IoT platforms, we have not found any holistic approach similar to the SecKit that considers in an integrated way of identities, context, risk, trust, and complex policy rules. Therefore, we present, in this section, existing approaches for trust management of IoT scenarios and for specification and enforcement of security policies in IoT and Body Area Networks (BAN) that are not as comprehensive as the SecKit.

The approaches that more similarly to ours consider the problem of IoT security and privacy in a global way, even if more oriented to the Machine-to-Machine (M2M) definition, are the standardization attempts proposed by the Open Mobile Alliance (OMA) and by the European Telecommunications Standards Institute (ETSI) Technical Committee for M2M. The lightweight M2M (LWM2M) protocol by OMA is designed for the remote control of M2M devices and applications. Like our toolkit, LWM2M addresses the problem related to access control but without the specification of complex policies and making extensive use of cryptography. In addition, it supports Digital Rights Management (DRM) [22, 8].

Similarly, ETSI proposes an architecture for M2M applications where message integrity and authentication of both messages and users are the main addressed problems [28, 29], resulting in simple access control mechanism if compared to our solution that, besides specification of complex rules through a policy language, supports also context changes.

The basic motivation behind the need of new approaches for IoT security is the limits shown by traditional mechanisms like ABAC (Attribute-based access control) and RBAC (Role-based access control), not scalable enough for IoT scenarios where a huge amount of data are processed and exchanged. Those limits are described already in Reference 5, where a Capability-based access control (CapBAC) is proposed. Although this solution supports rights delegation and customizable access rules, our toolkit provides also policies specification. Hernández-Ramos *et al.* [7] build up on the capability-based approach and propose a distributed version using capability tokens for CoAP Resources signed with the Elliptic Curve Digital Signature Algorithm (ECDSA) in order to ensure end-to-end authentication, integrity and non-repudiation. This solution allows IoT devices to take policy decisions without contacting a central PDP component, which is sufficient with policies that do not require external information. The authors also show a solution where a central PDP component is used to support context-based policies but no details are presented about the language and structure of these policies. Considering that, in order to prevent unauthorized access, revocation of tokens is needed in the capability-based solutions, the IoT devices always have to contact a revocation authority to verify if tokens are still valid. Furthermore, policies with complex context-based or trust-based conditions would require more processing power from IoT devices and queries to external information (e.g., context managers) in order to decide if the access should be allowed or not. A subset of our policy language with simple conditions and improved expressiveness could be used in capability tokens still with the benefit of allowing delays and modifications in addition to allow or deny access to an IoT device. An important result of this work is the performance analysis in very constrained devices with respect to the use of ECDSA.

The examples above show the importance of rule-based languages for security of IoT platforms. An existing applicable option is also the eXtensible Access Control Markup Language (XACML) policy language [26], which supports authorization request and response messages using attribute assertion operators. These operators are only propositional, while our toolkit supports also temporal and cardinality operators, thus allowing for the creation of policies not expressible with XACML.

A policy language more similar to ours is proposed, in the context of BAN, by Keoh *et al.* [12], who provide an implementation of a PDP component for the Ponder2 policy language in the TinyOS platform. From a point of view of the expressiveness language, Ponder2 supports Event–Condition–Action (ECA) rules with simple propositional conditions without support for context-based, trust-based, temporal, and cardinality operators. The authors present performance results for policy evaluation in the order of 50 ms for simple policies with one event and a simple action without taking into consideration the distribution of components. Our performance results are not comparable, since our policy language is more expressive and we consider the distribution of PEPs and PDPs, without evaluating policies directly in the constrained IoT devices.

Apart from access control and policy language, a crucial problem in IoT security is the level of *trust* users have in the system. Trusting a system means expecting a correct and fair behavior from it, and this expectancy comes directly from the reputation of the system itself, which is determined by all the users. The solutions proposed in the literature are basically inspired by some mechanisms originally designed for peer-to-peer and grid systems, where the problems of trust and reputation were faced already many years ago basically with the use of trust ratings, risk evaluation, and short-term behavior prediction [27, 13]. Later on, concepts also originally developed for semantic web and based on ontologies to find application in IoT [6], even if mainly targeted to the general problems of heterogeneity and scalability. Starting from all these concepts, the solutions for IoT add emphasis also on other important aspects like, for example, the social interaction, as proposed in Reference 1. Here, IoT devices are considered human operated, so the social aspect is central and the trust evaluation is based on properties like honesty, cooperativeness, and community interest. Social interactions trigger events in the trust management protocol, which analyses both direct observations and indirect recommendations. Although trust evaluation is performed by each node only on devices of its interest, the results show that the effectiveness of the protocol, tested on a trust-based service composition application, is comparable to protocols with global knowledge.

Other trust management proposals do not consider the problem in an integrated way, that is, there is no strong link with the model of the system. For example, in Reference 33, the association between trust relationship and devices in the system is unclear, making difficult to determine the influence of the first one on activities of a real scenario. Our toolkit does not address specific challenges alone, as done with the ontologies-based approaches, but it is characterized by a strong association between trust and system models along with the integration of a security policy language.

5.3   IoT framework

The work proposed in this chapter derives from the iCore Project [31], where Services in the IoT framework are refined into CVOs and VOs, which are digital representations of RWO. A complex object such as a Smart Home can be represented as a CVO consisting of a set of VOs corresponding to sensors (e.g., temperature and movement) and a communication system (e.g., Wi-Fi network). The resulting complete IoT system consisting of CVOs and VOs provides access to its capabilities by means of Services.

Figure 5.1 presents a high-level overview of the iCore Framework layers and security management components, for a detailed overview we refer the reader to Reference 20. An user, which includes both a human being and an application, can initiate a request to the iCore Framework through a Service Request (at the top left of Figure 5.1), which is analyzed and semantically interpreted in the service layer of the iCore Framework using an SPARQL query [32]. On the basis of the profile of the user and preferences, the iCore Framework selects, using a semantic query, the appropriate Service, CVOs and VOs to fulfill the request. The selection is based on the current status of the Real-World Knowledge (RWK) as well as System Knowledge (SK) databases, which are dynamically adapted. The RWK maps the current status of the real world, while the SK is updated with the internal status and resources of the iCore Framework.

Services, CVOs or VOs are instantiated using predefined templates that provide semantically the best match to the requests by the user. The VO and CVO layers follow a common pattern: a factory component instantiates CVOs and VOs from templates, and a management unit updates the registry with information about existing instances that run in a container. The container components in the respective layers are the execution environment of CVOs and VOs. The physical layer consists of the

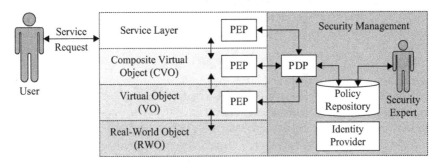

Figure 5.1   iCore architecture

RWO themselves with custom gateway components that act as wrappers of the RWO functionality for integration in the iCore Framework.

The Security Management layer consists of the cross-layer components provided by the *SecKit* for authentication and security policy evaluation, which are integrated in the iCore Framework. The Identity Provider component is responsible to authenticate users to access the framework, which can be achieved using different technical solutions. The specification of policies referencing identities is done using an abstract identity design model that can be mapped to the specific technical choice.

The Policy Repository stores the defined security policies and templates specified by a Security Expert. The PDP is in charge of verifying the conformance of the requested Services, CVOs, and VOs activities with respect to the instantiated policy rule templates retrieved from the Policy Repository. The evaluation of the policies is done using events signaled by the PEPs deployed at the different layers of the iCore Framework. The PDP receives the events from the PEPs, checks if the active policy rule instances are triggered, and returns enforcement actions to the PEPs.

5.4   Methodology for security engineering in IoT

In this section, we describe a model-based methodology for security engineering in IoT systems, which is supported by a toolkit named the *Model-based Security Toolkit* (SecKit). The main objective of our approach is to address the IoT challenges for security and privacy, more specifically addressing the challenges of support for dynamic context, trust management, digital divide, control of the data flow from IoT devices, control of actions of IoT actuators, and anonymization of data. A more extensive description of these challenges is described in Reference 20.

In a world where IoT objects are more and more interacting and exchanging data, it is extremely important to be able to define and impose "rules of conduct" through mechanisms allowing to identify the most suitable security policies to be applied in a given scenario, to define the level of *trust* of the counterpart, and to regulate the information flows. The SecKit aims at achieving these objectives supporting integrated policy specification and enforcement at the Service, CVO, and VO layers of the iCore Framework. The SecKit foundation is a collection of metamodels that provides the basis for security engineering tooling, add-ons, runtime components, and extensions to address security, data protection, trust, and privacy requirements.

In contrast to other approaches, the SecKit precisely specifies the relation between the security concepts and other security-relevant system concepts. Furthermore, the SecKit metamodels can be used as an abstract reference for conceptual agreements between different domains contributing to the interoperability alignment between them. The defined metamodels include Data, Time, Identity, Role, Context, Structure, Behavior, Risk, Trust, and Rule metamodels implemented using the Eclipse Modeling Framework (EMF) [3] to support the specification of types, instantiations, and instances of the various concepts.

The following lists summarizes the steps of our methodology for security engineering in IoT systems and the respective models and issues analyzed in each step:

1. **Structure and behavior**: the structural model specifies the entities and interaction point mechanisms between these entities representing the communication mechanisms. Each entity defined in the structural model is assigned a behavior, specifying the activities consisting of actions and interactions between entities.
2. **Data and identities**: activities in the behavior model define the data instances established by them and identities used for authentication. Types of each data and identity instantiation must be defined in this step.
3. **Business roles**: specify role types and role hierarchy with subtypes.
4. **Context information and situations**: specifies context information, context situations, and Quality-of-Context attributes.
5. **Risk analysis**: specifies assets, vulnerabilities, threats, risk, and countermeasures mapped to trust relationships or security policy rules.
6. **Trust relationships**: specifies trust relationships related to specific trust aspects representing dependencies with external stakeholders.
7. **Security policy rules**: specifies abstract ECA rule templates and configurations that instantiate the rule templates when specific conditions are satisfied. Additionally, in this step, the different technologies used in the system and the strategy for implementation of enforcement monitors and the required levels of abstraction should be identified.

SecKit can be used to specify and enforce policy rules for anonymization, confidentiality, data retention (e.g., delete data in 30 days), user consent, access control, non-repudiation, and trust management. The policy rules in SecKit, consisting of authorizations and obligations, are specified as ECA enforcement rules. These rules use as a reference set of inter-related *design models*, conforming to their respective metamodels, representing the different aspects of the IoT system. These design models are used as input by the runtime models and components in the SecKit enforcement platform, enabling monitoring of ECA rules and execution of enforcement behavior.

The association of the ECA rules with these design models allows for checking of policy consistency with respect to the IoT system design models and the precise identification of PEPs. All the metamodels introduced in the following subsections are connected to the policy rule language described in the last subsection by means of events, event patterns, and specific pattern matching operators. In the following subsections, we describe each step of our methodology including the dependencies between the models and important issues that should be considered during each step. Screenshots of the SecKit Graphical User Interface (GUI) implementation for all the different models are presented in the case study in Section 5.6.

5.4.1 Structure and behavior models

The SecKit structure and behavior metamodels are inspired by an existing generic design language to represent the architecture of a distributed system across application

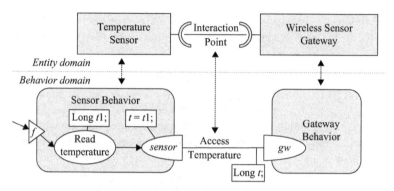

Figure 5.2   Entity and behavior domains

domains and successive levels of abstraction called Interaction System Design Language (ISDL) [24]. The system design in ISDL is divided into two domains named *entity domain* and *behavior domain*, with an *assignment relation* between entities and behaviors.

In the entity domain, the designer specifies entities and interaction points between entities representing the communication mechanisms. In the behavior, domain the behavior of each of the entities is detailed including actions, interactions, causality relations, and information attributes. The idea behind the SecKit models inspired by ISDL is to provide a minimal set of concepts that supports the design of the IoT Services, CVOs, and VOs from the iCore Framework described in Section 5.3. Furthermore, one important feature of ISDL that justifies our choice for this language is the support for refinement relations, which has being applied in previous work to support the automated refinement of security policy rules [17].

Figure 5.2 shows an example structure and behavior design models at one arbitrary abstraction level. In this example, the *Temperature Sensor* entity interacts with the *WirelessSensorGateway* through an *Interaction Point*. The interaction type and the information exchanged are depicted in the behavior model, which in this example is the *Access Temperature* interaction, which exchanges the temperature measurement (t). The contribution of each behavior type to the interaction is depicted by a half circle representing the role of each behavior in the interaction, in this example the roles are *sensor* and *gw*, respectively.

Figure 5.3 illustrates the mapping of VOs, CVOs, and Services to behavior types and instantiations. VOs are mapped to monolithic behaviors while CVOs are mapped to composite behaviors that contain other behavior instantiations, which could be a VO or recursively another CVO as well. In the two Service Layer users, which may be human users or other IoT systems, access the CVO functionalities through service interactions.

In our behavior metamodel, we define the possible activity types (actions and interactions), instantiations, and instances. Activities are data consumers and producers that are enabled to be executed by means of causality relations. Here, we consider

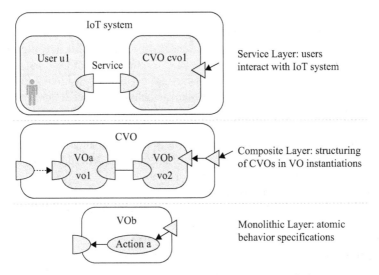

Figure 5.3   Structuring of VO, CVO, and Service behaviors

only the specification of behaviors and activities that are relevant for specification of security policy rules.

An interaction type is defined with a name, a set of interaction contribution types, a set of data instantiations, and a set of identity instantiations representing the values established after an occurrence of an interaction instance of this type. We also define interaction instantiations contained in behavior types at design time, and interaction instances contained in behavior instances at runtime. An interaction contribution type specifies a behavior role and a set of data and identity instantiation names that are the input of the specific role for the interaction type they participate in. For example, in Figure 5.2, the *Smart Home Behavior* assumes the *home* role and contributes with the *hr* data to the *Access Heart Rate* interaction. Interaction types can be mapped to service specifications, where typical roles are providers and consumers. However, in our model, abstract interactions may define more than two roles in an interaction.

Interaction contribution types are instantiated by behavior types and behavior instantiations. Interaction contribution instantiations of a behavior type define possible contributions of a type that are instantiated for all instantiations of the specific behavior type. An interaction instantiation connects two or more interaction contribution instantiations to define a concrete interaction possibility between behaviors. In our policy rule language, we define activity events and event patterns that match activities defined in the behavior model, and may also include patterns to match the data and identity produced by the activity. For example, a security policy rule can be specified to deny all network interactions of a specific behavior instance when there is a possibility that personal data may be exchanged.

In addition to interaction instantiations, behavior types also specify the contained action instantiations, flow point instantiations, and the causality relations between them. Even though we include in our metamodels the causality relations are not needed to support the specification of policy rules. Causality relations are useful to generate executable behavior specifications, for simulation of behavior execution, and to support information flow analysis, which is part of our future work.

The security policy rules proposed by us are evaluated considering events that represent the execution of activities defined in our behavior model. In order to support the specification of these policy rules, we define behavior and activity pattern matching relations. Interaction patterns match an interaction of a specific instantiation, type, and that establish specific data and identity results. In addition, an interaction can also be matched considering the pattern of interaction contributions participating in the interaction. For example, a privacy protection policy rule can be defined to deny a *data request interaction* (type) defined between a weather station and a cloud service (instantiation) if the identity of the station owner is provided by the weather station (contribution).

In the matching of entities and behaviors, we also allow the use of trust relationship patterns, introduced in Section 5.4.6. Patterns can be specified using the typed *any* wildcard to match all behavior types, instantiations, and instances. This flexibility in our pattern matching approach allows policy rules to be specified that apply to all interactions contained in an IoT system, where a specific smart home interacts with any other entity. Examples of the structure and behavior design model GUI are shown in our case study in Section 5.5.

5.4.2 Data and identities

After specifying the entities and the respective behaviors and activities the data types used should be also specified. Activities in the behavior model declare named instantiations of the data types at design time, while data instances are created at runtime. The set of data types is partitioned in primitive types, defined using the primitive data types of EMF, and composite data types. Both composite and primitive data types are defined in the context of a data type package.

Composite data types define attributes of a primitive data type but may also refer recursively other composite data types. Composite types may also be defined as a subtype of other composite types. Therefore, we also include a constraint to forbid a composite type to be indirectly a subclass of itself, which is in alignment with the specifications of the EMF metamodel.

In our security policy rule language patterns can be specified in policy rules to match data instances at runtime in the rule conditions. Patterns may include variables in order to enable configurable security policy rule templates. Our strategy is to define named variable declaration that can be used in place of the respective type in the pattern specification. We follow the same strategy in the specification of patterns for identities, context situations, roles, structural, and behavioral elements. We also support in our implementation the interpretation of data values as regular expressions, XPath expressions, or static literal patterns. We allow the specification of data patterns that match more than one data instance using the reserved typed wildcard *any* for data type and instantiation names.

In our metamodel, we relate data instantiations and data instances with a set of data concepts. An ontology of data concepts is useful in the specification of abstract security policy rules that apply to a concept independently of the system activity that handles the data. The definition of a data ontology, tracking static information flow analysis at design time, and dynamic information flow tracking at runtime are out of the scope of this chapter.

Data types are instantiated in the specification of identity types, context information types, and activity types to represent their data input and output. The data model also supports inheritance, which enables specification of security policy rules for abstract data types. For example, an enforcement template defined for the *PersonProfile* is also applicable for a *SocialNetworkProfile* if the latter is a subtype of the former. Data types can be imported from an ECore model specified using the standard Eclipse EMF tooling.

The identity model specifies the identity types and their respective identity attributes. Our model abstracts from specific technical details of the adopted identity management solutions, it only provides a high-level description of the identity types and attributes that can be used in the specification of security policy rules. Our model of identity types follow a simple attribute-based approach, where the subject name and identity attributes are part of the data instantiation set.

An identity type is defined by a name, a mandatory subject name of the *string* type, and a set of identity attribute declarations. Identity instantiations are used in conjunction with data instantiations to model the results of activities in the behavior model. An identity instance includes an identifier of the identity and one of the identity issuer, allowing for self-signed and third party certified identities.

Identity patterns are also supported for both types of identities in our security policy rule language. An identity instance matches an identity pattern if the instantiation is a match, if the subject is a match, and if all the attributes are a match for the pattern. Furthermore, for certified identity patterns the identity issuer may also be matched. Similar to data patterns, we support the specification of identity patterns that match more than one identity type, using the *any* wildcard for identity type and instantiation names.

5.4.3   Business roles

The role model specifies the role types and the role hierarchy with a possible inheritance of membership from role types. Identities are assigned to role types and a sub-role relation maps a role type to a parent role, with the added constraint to prevent cycles in the role hierarchy definition. A role pattern is trivially the specification of a type that should match a specific role, or indirectly a parent role, and an identity pattern that should be contained in the matching role type hierarchy.

In our security policy rule language role patterns can be specified in policy rules to match role membership at runtime in the rule conditions. A role pattern can be used to allow or deny the execution of an activity depending on the assigned role. For example, *Doctor* and *Nurse* are both sub-roles of the *HealthProfessional* role, where entities assigned to the child roles are also members of the parent role.

The assignment of identities to role types is done in the runtime model specification in the SecKit.

5.4.4   Context information and situations

A Context Information is a simple type of information about one entity that is acquired at a particular moment in time, and Context Situation is a complex type that models a specific condition that begins and finishes at specific moments in time [2]. For example, the Global Positioning System (GPS) location is an example of a context information type, while *Fever* and *In One Kilometer Range* are examples of situations where a patient has a temperature above 37 °C, or a target entity has a set of nearby entities not further than 1 km away. Patient and target entity are the roles of the different entities in that specific situation.

A context information type is defined as a mapping from a name to a data instantiation. A context situation type is defined as a mapping from a name to a set of situation role names. We allow in our security policy rule language situation events and situation event patterns to detect the beginning and end of a context situation. In our trust model, we define the personal and situational trust types that specify trust relationships valid for a set of entities in a specific situation. With our context metamodel, we address the challenge *Support for Dynamic Context* identified in the list of challenges introduced in Section 5.4.

5.4.5   Risk analysis

In the SecKit methodology, trust relationships and security policy rules are motivated by a risk model that clarifies the concrete known harm of using IoT systems. The risk model should be specific, clear, and well understood by the IoT system users. Our risk modeling approach is an extension of the CORAS methodology for risk assessment [14].

The main element of the risk model is a Threat Scenario, which depicts with a level of probability how threats can cause incidents by means of exploiting existing vulnerabilities. Incidents have negative consequences on assets owned by stakeholders with a level of severity, for example, the result of a leak of presence information may trigger a privacy violation. The resulting risk calculation of a threat scenario is done considering the probability versus the severity of the negative consequences caused by the resulting incidents. One threat scenario may enable other contained threat scenarios to happen, given that additional vulnerabilities are present. IoT system users can maintain a history of incidents and exchange information with respect to occurrence of threat scenarios enabling empirical statistical analysis of risks.

Threats, assets, and vulnerabilities are associated with the system elements defined in the behavior and entity models and are not specified in isolation. This association provides traceability of risks, supports structured analysis, and allows more concrete countermeasures to be adopted. Each IoT system user may decide to adopt the indicated countermeasures if the risk level is above a certain personal threshold. Countermeasures in the risk model can be associated with a set of rule templates that can be deployed and enforced to detect and react or prevent the threat scenario from happening. A second choice is to associate a countermeasure with a rule template that evaluates a minimum required level of trust in a certain entity that reduces the probability of the specific threat scenario happening.

For example, a threat scenario could be specified for an IoT device in a smart home that allows external entities deliberate access to information, and, as a consequence, these entities can infer if the homeowner is at home or not. This threat scenario enables a second threat scenario where thieves rob the smart home when the owner is not at home. One possible countermeasure to mitigate the enabling threat scenario is to instantiate a security policy rule template that control external access to the IoT device. More abstract countermeasures can also be specified, for example, to deploy an effective alarm system that prevents thieves from robbing the citizens' home.

5.4.6   *Trust relationships*

In our trust model, we define trust as relationship between a *Trustor* and a *Trustee* [10]. We define trust as *the measurement of the belief from a trusting party point of view (trustor) with respect to a trusted party (trustee) focused on a specific trust aspect that possibly implies a benefit or a risk* [21]. For example, *Bob* (Trustor) may trust to a *high degree* (measurement) his friend *Alice* (Trustee) concerning its *punctuality for appointments* (trust aspect). The benefit or risk implication is only present when *Bob* accepts to depend on *Alice* for some activity and may be impaired by inaccuracies, for example, to depend on her to arrive on time for an important meeting. A trust relationship is defined from a trustor perspective (referenced using an identity) for a specific trustee scope, it has an abstract likelihood measurement, and it is defined for a specific trust aspect.

Our model supports the specification and reasoning about different *types* of trust relationships and *recommendations*. Trust relationships can be specified directly based on a trustor's arbitrary opinion or considering previously observed positive/belief or negative/disbelief evidence about a specific trust aspect. Indirect trust is specified using the concept of fusion operators that combine a set of trust relationships that match a specific pattern. For example, a reputation trust value can be defined by a combination of recommendations from many entities or a community. Positive and negative evidence are translated for each trust aspect to an opinion using the formula illustrated in Reference 21 (p. 91).

Different likelihood measurement approaches can be adopted: in our methodology, we propose to use the Subjective Logic (SL) [11], which is a probabilistic logic capable of explicitly expressing uncertainty about the probability values. In SL, an opinion is represented by the belief (b), disbelief (d), and uncertainty (u) where $b + d + u = 1$. We adopt a discrete mapping of an SL opinion to a set of five *trustworthiness* values using the following semantics: *when the uncertainty is lower than the belief/disbelief this represents a very trustworthy/untrustworthy opinion* [21].

The objective of a trust aspect definition is to have a precise specification of what is the intended meaning of a trust relationship, including issues related to the provisioning of trust recommendations, structural, behavior, data, and identity qualities. The following list describes the meaning and example of the trust aspect subtypes:

- The *Recommendation Quality* aspect is associated to an entity to provide trust recommendations. For example, someone may be trusted to recommend a good car mechanic but not to recommend a restaurant.
- The *Structural Quality* aspect relates to properties of elements in the entity domain of our system model that are independent of the behavior like, for example, physical tamper resistance. This trust aspect is related to entity types (e.g., Weather Station) and interaction point types (e.g., Wi-Fi).
- The *Behavior Quality* aspect is the quality aspect of an entity behavior that can be specified considering the quality of the service (e.g., response time), experience (e.g., usability), or protection (e.g., privacy enforcement). This trust aspect is related to activity types (e.g., *Update firmware*) and interaction contribution roles (e.g., *manufacturer*).
- The *Data Quality* aspect is associated to the capability of an entity to provide data according to a specified quality level. In an IoT scenario, this can be associated to the precision of sensor information (e.g., temperature).
- The *Identity Quality* aspect is a specialized type of *Data Quality* trust related to the level of assurance (LoA) of the identities provided by an entity. This trust aspect is usually associated to identity providers and the quality of identity attributes used in the identities issued by them (e.g., attributes verified face-to-face).

Trust aspects may overlap when trust relationships are specified for data or identity provisioning services, which in this case are assigned, respectively, to data and identity quality trust aspects and the specified interaction type may be omitted. Some trust models associate trust relationships with an isolated or combined measurement of the concepts of honesty, credibility, reputation, usability, competence, fit for purpose, reliability, and quality [15, 25]. Our trust model captures precisely all these different concepts with the definition of the different trust types, recommendations, and trust aspects. With our trust metamodel, we address the challenge *Support for Trust Management* identified in the list of challenges introduced in Section 5.4.

From a methodological point of view, security engineers should analyze the interactions defined in the behavior model and identify dependencies considering the roles participating in each interaction type. Furthermore, in specific instantiations of the interaction types additional trust relationships may be included, for example, in specific interactions a bigger set of trust relationships may be necessary to ensure data quality, secure communication mechanisms, and so on. Trust relationships are then referenced in security policy rules using trust patterns in the specification of precise trust requirements that should be satisfied at runtime.

A trust pattern allows matching of trust relationships and recommendations in our security policy rules considering a specific trust aspect, degree, trustor, trustee scope, time interval, and recommenders' identities. Using this flexible model of combination of trust based on patterns, it is possible to infer derived trust relationships from different aspects: for example, privacy enforcement and identity provisioning trust could be combined in a more general service provisioning trust relationship. The trust patterns are evaluated considering the available trust relationships in the trust database.

Trust assessments are useful in our policy rule language to define conditions, for example, allowing a specific activity only if a minimum likelihood measurement is met. We define three assessment operators, namely *exactly*, *atLeast*, and *atMost*, which are evaluated by the *evalTrust* relation between a likelihood measurement and a trust query. Trust queries are evaluated using the *query* relation, using the previously defined pattern matching relations for trust relationships and recommendations, consider the trust relationships from one trustor point of view or the trust recommendations received from other trustors' point of views. We only support one concrete example of a fusion operator in our policy rule language, the *Consensus* trust query element, which is mapped to the SL consensus operator [11] and implemented in the SL API [23]. The result of this operator is increased uncertainty if the combined likelihood measurement values in the received trust recommendations contradict each other.

5.4.7   Security policy rules

The specification of policy rules in SecKit is done using policy rule templates that must be explicitly configured. A policy rule template follows an ECA structure, evaluated over a discrete trace of sets of events with the following semantics: when the trigger event (E) is observed, and the condition (C) evaluates to true, the action (A) is executed. Rule templates refer to the design models of the system data, identity, role, context, structure, behavior, and trust. In this subsection, we show the formalization of events, event patterns, the operators supported in the condition of a policy rule, and the formalization of policy rule templates and configurations.

Events in the SecKit represent context situation changes and activities (actions and interactions) between VOs, CVOs, and Services in the iCore Framework. We model the start of an activity, ongoing activities, and the completion of an activity with the event indexes: start, ongoing, completed. To support enforcement of usage control policies including authorization decisions, we model *tentative* and *actual* events. A tentative event is generated when an activity is ready to be started by the iCore Framework but has not started yet, giving the opportunity for the execution of enforcement actions to allow or deny the execution of the activity.

Event modalities are only used for activity events because context situations are only detected and no enforcement can take place to allow or deny situations, they are simply observed. With the combination of the *ongoing* index with a *tentative* modality, we are able to model ongoing activities that can be stopped by the policy rules. A discrete event trace is represented as a mapping of time steps $\mathbb{N}$ in a *TimeStepWindow*, which contains a set of observed events.

We also support events to represent the life cycle (instantiation and disposal) of system entities, behaviors, identities, context information, and data. Due to space restrictions we do not show the detailed formalization of these types of events, we focus on the specification of interaction events. The formalization of actions, flow points, context situations, and life cycle events are done in a similar way. Interaction events reference instances of the respective interaction instance, which contains details about the activity being executed.

Event patterns are specified for all event types and also for a special event that denotes the end of a discrete time step named *TimestepEvent*. Interaction event patterns include patterns for the interactions contributions, data and identity established by the interaction, behavior where the activity is contained, and for the entity that executes the behavior. Interaction contribution patterns also include behavior and entity patterns to match the participants of the interaction. In our policy rules, we define conditions based on the occurrence of events, and to specify this conditions we define event pattern types with a $\models_{pattern}$ satisfaction relation.

The condition part of a rule template is a formula Φ, which consists of constants, event patterns, trust patterns, propositional, temporal, and cardinality operators. The abstract syntax of a formula is defined as:

$$\Phi ::= \underline{true} \mid \underline{false} \mid EventPattern \mid TrustAssessment$$
$$\underline{not}\langle\!\langle \Phi \rangle\!\rangle \mid \underline{or}\langle\!\langle \Phi \times \Phi \rangle\!\rangle \mid \underline{and}\langle\!\langle \Phi \times \Phi \rangle\!\rangle \mid \underline{implies}\langle\!\langle \Phi \times \Phi \rangle\!\rangle \mid$$
$$\underline{always}\langle\!\langle \Phi \rangle\!\rangle \mid \underline{eventually}\langle\!\langle \Phi \rangle\!\rangle \mid \underline{since}\langle\!\langle \Phi \times \Phi \rangle\!\rangle \mid$$
$$\underline{before}\langle\!\langle \mathbb{N} \times \Phi \rangle\!\rangle \mid \underline{within}\langle\!\langle \mathbb{N} \times \Phi \rangle\!\rangle \mid \underline{during}\langle\!\langle \mathbb{N} \times \Phi \rangle\!\rangle \mid$$
$$\underline{repSince}\langle\!\langle \mathbb{N} \times \Phi \times \Phi \rangle\!\rangle \mid \underline{repLim}\langle\!\langle \mathbb{N} \times \mathbb{N} \times \mathbb{N} \times \Phi \rangle\!\rangle$$

Informally, the semantics of the temporal operators is: for *always* the operand must have been true in all previous time steps; for *eventually* the operand must have been true at least once in all previous time steps; for *since* first operand must have been true in all time steps since the second operand was true, or the second operand has always been true (weak since from Linear Temporal Logic (LTL)); for *before*, *within*, and *during* the formula operand must have been true at a specific number of time steps ago, at least once for a bounded number of previous time steps, or during all bounded number of previous time steps; for *repSince* is the same as since with an additional upper bound number of times that the first formula may have been true; for *repLim* the formula must have been true during a bounded number of previous time steps according to a lower and upper bound for the number of occurrences. The formal semantics of the temporal and cardinality operators is based on past time LTL.

The specification of authorization and obligation policy rules is done in SecKit using a model containing rule templates (*RuleTemplate*) that must be explicitly instantiated using rule template configurations (*TemplateConfiguration*). Templates are parametrized with variables that are instantiated by the template configuration. When the rule is configured it is also possible to specify when the rule should be disposed, after it is triggered the first time, never, or also when a particular Event–Condition is observed. By introducing templates we address the challenge *Support for the Digital Divide* identified in the list of challenges introduced in Section 5.4, allowing to less knowledgeable users the selection of existing templates instead of specifying their own.

[*VariableName*]
AuthorizationResponse == *Allow, Deny*
Modify == *DataInst → NewValue*
Delay == *TimeAmount × TimeUnit*
Enforcement == *AuthorizationResponse* × $\mathbb{P}$ *Modify* × *Delay*
RuleTemplate ::=
    $\mathbb{P}$ *VarDecl* × *EventPattern* × Φ×
    Enforcement × $\mathbb{P}$ *TrustUpdate* × $\mathbb{P}$ *BehaviorInst*
CombiningAlgorithm ::= *FirstApplicable* | *AllowOverrides* | *DenyOverrides*
TemplateConfiguration ::=
    MonolithicConfiguration⟨⟨$\mathbb{P}$ (*VarAssign* × *RuleTemplate* × *RuleTemplate*)⟩⟩ |
    RecursiveConfiguration⟨⟨*CombiningAlgorithm* × *seqTemplateConfiguration*⟩⟩
RulePackage == $\mathbb{P}$ *RuleTemplate* × $\mathbb{P}$ *TemplateConfiguration*
∀ *e : Event*; *ep*1, *ep*2 : *EventPattern* •
    rulenestedrule ⇔ (*ep*1 *matches e*) ∧ (*ep*2 *matches e*)

The event part of a rule template is called the *trigger event*, and is an event pattern matching operator that is also supported in the condition part. An event includes the details about the activity being executed such as identity and roles of the entities performing the activity, and information attributes produced by the activity itself. For action events we can trigger rules considering the entity executing the action, for interaction events we can observe who are the interaction participants and perform the respective enforcement. We also support patterns of events capturing context information and context situation changes, and life cycle events for entity, behavior, and data types.

The action part of an enforcement template consists of an enforcement and an execution part. The enforcement part refers to the trigger event of the rule template and may allow or deny the execution of the respective activity. If the activity is allowed, it is also possible to specify an optional modification or delay of the activity execution, for example, anonymizing activity data before the activity takes place. The anonymization could be the simple modification of a *name* identity attribute to an empty string. The execution part of an enforcement template may trigger the execution of additional activities, for example, notifications or logging of information. With the addition of these different enforcement and execution options in our policy rules, we address the challenges *Control of the data flow from IoT Device, Control of the actions of IoT actuators*, and *Anonymization of data* identified in the list of challenges introduced in Section 5.4.

In addition to enforcement templates, we also support instantiation templates, disposal templates, and condition templates, in order to maximize reuse of enforcement rules and to allow efficient monitoring of parametrized templates. The action part of these templates implicitly instantiate and dispose other rule templates when the event and condition parts evaluates to true.

When composite enforcement templates are specified, a combining strategy must be also specified in case multiple rules are triggered resulting in a conflicting decision similar to the XACML approach [26], such as allow overrides, deny overrides, or first applicable rule. Furthermore, child/contained templates must specify a refined trigger event in relation to its parent, which informally means that the trigger pattern of child templates inherit the patterns specified in the trigger of the parent rules. The general idea is to support the reuse of the instantiation template to other enforcement templates without loss of generality.

At runtime, all template configurations will be instantiated using the variable assignments defined, resulting in a set of rules. Our combining algorithm always select the modification or delay to apply considering the enforcement chosen according to the strategy definition. The semantics of the combining algorithm is recursive, where contained rules that contain nested rules themselves are first evaluated and the resulting enforcement is combined with the parent rule for a unique enforcement result in a rule hierarchy.

In the SecKit, we adopt a general-purpose runtime event signaling and policy evaluation strategy. Rules are instantiated by a *Rule Template Configuration* at a particular moment in time, and an enforcement rule instance is created. Each rule is configured with a *time step size*, which indicates the granularity of the enforcement and events are observed for each *time step window* according to this granularity. Our PDP component is a rule engine that keeps trace of all time step granularities for the existing rule instances, observes the events, and evaluates the rules at the end of each time step window.

The PDP monitoring approach does not store all events observed, only the events in the current and previous time step windows are kept following an approach similar to Meredith *et al.* [16]. It uses simple three-value Boolean states for all operators, counters for the cardinality operators, and a circular buffer of n Boolean states for the time-bounded operators supported in our language like *before*(n, φ), which requires only the storage of the truth value of φ for the n previous time step windows.

The generation of efficient runtime monitors for parametrized specifications is a nontrivial task [16], which is the case of rule templates. Monitoring of parametrized templates is difficult to be realized because the number of parameter bindings can be very large, and there are only domain-specific solutions to handle this problem. In our approach, we specify explicitly the instantiation and disposal of rule templates with variables to enable the efficient generation of monitors. By using the explicit instantiation of parametrized templates, we avoid the overhead of monitoring all possible combinations of observed variable values, which may be practical for a small set of variable values, but it easily becomes intractable due to the large number of combinations [9].

In order to be useful at concrete implementation scenarios, the SecKit must be extended with technology-specific runtime monitoring components. In the iCore Project, we provide one extension to support monitoring and enforcement of policies for an MQTT broker, which is the technology in our case study to support communication between VOs and CVOs. In general, PEP components should be able to intercepts the messages exchanged between the system entities, notifies these messages as events in the SecKit, and optionally receives and enforcement action to be executed (e.g., Allow, Deny, Modify, etc.). Details about the MQTT enforcement component are described in Section 5.6.

5.5   Smart business case study: Cold Chain Monitoring

In this section, we describe the smart business case study proposed in the iCore Project. The main goal is to support *Cold Chain Monitoring* (CCM), which consists of the real-time monitoring of the transport conditions of perishable goods. More specifically, the CCM objectives are the following:

(a)   to provide real-time alerts and early warnings during transport in case the required transport criteria are not met;
(b)   to issue acceptance reports for the end users receiving the perishable goods;
(c)   to control automatically the storage and transport conditions of goods.

The iCore Framework described in Section 5.3 provides an excellent ecosystem to link the mesh of ICTs and intelligently manage them, according to delivery specifications, and based on the current (and changing) environmental context. We show in Section 5.6 how the high-level information sharing and data confidentiality requirements introduced in the description of this case study are addressed using the SecKit methodology.

5.5.1   Overview and motivation

In a cold chain logistics system, goods are transported from suppliers to retailers via a mesh of warehouses connected by road, air, or sea transportation in between. The main end-user concern is the lack of insight in the storage and transport conditions of goods between suppliers and consumers. To address this concern in this case study, a fine-grained Information and Communications Technology (ICT) monitoring system (wireless sensor network – WSN) is used.

An important element in CCM is acceptance reports showing if the transported goods were kept within the specified temperature tolerances. These reports are useful, for example, for a retailer to know if he can accept a shipment of temperature-sensitive medicines, or for transport operators to know when they can avert a claim of spoiled goods. In case of violation of storage and transportation conditions, the parties responsible for the goods also want to be able to act as soon as possible to reduce product spoilage (and associated claims). In addition to acceptance reports, suppliers/retailers want to track and know in advance when to expect a delivery of goods.

Food and pharmaceutical goods are exposed to increasingly long and complex supply chains with many dangers of poor temperature control, delays, and physical mishandling. As a result, these products suffer reduced quality and safety, and inevitably increased waste. Besides the financial loss suffered from waste, low-quality products often do not meet consumer expectations which result in an slow overall decay of high value chains. Another negative impact of food losses caused by imperfect chains is that of low environmental sustainability. Reducing the amount of lost and damaged perishable goods during transportation and storage during the life cycle of the product is a substantial global challenge.

It is essential that products being transported in a cold chain remain within the specified acceptance criteria. A violation of the acceptance criteria can harm the customer's health using the product. In addition to acceptance reports, another goal of this case study is to produce early alerts about possible temperature violations. One study found that opening the doors of a truck with frozen goods may lead to an air temperature increase of 24 °C within 2 min [30]. Therefore, to keep the products within the acceptance criteria a fast reaction in a matter of minutes is necessary.

Key to reducing these huge waste numbers is the implementation of CCM at all levels of the supply chain. The most significant challenge is usually not gathering the data but developing an integrated strategy that allows virtually everyone involved with supply chain processes to improve the use of the data for decision-making. The transport companies are interested in the adoption of these new technologies which can optimize the cold supply chain and can reduce the amount of lost and damaged perishable goods and can increase the quality of services offered to the customers.

Several technologies, methods, and processes are already current practice and have been implemented in the supply chain, such as modified atmosphere packaging, temperature control, sanitation processes, and CCM. At present, CCM is mainly performed by the use of data loggers and typically only during transport and not throughout the entire supply chain. In summary, the CCM case study proposed in iCore demonstrates the following:

- Acceptance reports: the system provides the end user of perishable goods (e.g., a retailer or pharmacist) a report at delivery of the products showing the transport and storage conditions. Based on these reports, the end user is able to decide whether to accept the shipment or not; the report can only be accessed after the shipment is delivered.
- Real-time alerts/early warnings: the system provides (early) warnings to trans-porters when the products are outside their optimum storage conditions. The system provides these warning to guarantee the quality of products throughout the supply chain and enables transporters to act before damage occurs. Access to the monitoring system at the warehouses is only allowed to the device owners and to the logistic companies responsible for the shipments being monitored.
- Automatic control of storage and transport conditions: since the CCM system has knowledge of the required conditions for products within a warehouse/truck, it can provide functionality to automatically control environmental conditions based on the requirements of products stored in (compartments of) warehouses or trucks. The control can be achieved by directly controlling heating, ventilating, and air conditioning (HVAC) systems in these warehouses and trucks.

5.5.2   Hardware components

This case study is mainly focused on the handling of temperature-sensitive products in the supply chain. Temperature is therefore one of the most important parameters to monitor, preferably close to the product itself. Ambient Systems, which was one of the iCore Project partners implementing this case study, has developed several temperature monitoring devices, which are used in this proof of concept. Figure 5.4 shows a temperature sensor embedded in the packaging of a product being transported.

Figure 5.4   Temperature sensor deployed close to temperature-sensitive product

The sensors are designed for mobile use and to travel with the product from manufacturing site to the end user. The sensors communicate wirelessly in the 2.4 GHz band and have enough memory to store months of sensor readings in case they are outside the coverage of WSN gateways (e.g., in a ship/an airplane). The sensors can be configured remotely and in general push temperature information once per 5 min (which is common practice in CCM). Optionally, the sensors can provide remaining shelf lifetime of the products they monitor, based on historical temperature conditions.

The following additional sensors are integrated:

- Open/close sensors are used to detect if doors in warehouses or trailers are opened or closed.
- Current clamps are used to detect the performance of HVAC systems. They measure the electrical current flowing through the appliance, which is an indirect measure for the load of the machine. This is used to learn the cooling/heating capacity.
- Indoor air quality sensor consists of relative air humidity, carbon dioxide, and temperature. This sensors are used, for example, to monitor certain fruits that are usually stored in high concentration of CO_2 rather than fresh air to improve their preservation by allowing them to retaining flavor and quality during transportation.

WSN infrastructure (such as routers and gateways) are "see-through" concerning sensor data (see Figure 5.5). The gateway communicates periodical via socket connections with the VO Back-end using a 3G connection. The gateways used in this proof of concept also provide location information (GPS), for example, to determine the location of the infrastructure (e.g., truck) it is mounted on.

Figure 5.5   A wireless sensor network gateway device

The scenario uses several sensors attached to the product within a cold chain container. By having the sensor data combined with the location information many different scenarios can be monitored. The cold chain dispatcher may decide between different cold chain warehouses nearby, or may choose to use a container-cooling unit that is only able to keep the temperatures, but not cooling down warm products. We also implemented additional rules in the scenario to improve the monitoring accuracy and capabilities, for example:

(a)   if one sensor differs from the other sensors within the container the container door may have been accidentally opened;
(b)   by determining the location, the outside temperature forecast can influence the sensor value in approaching the acceptance criteria.

5.5.3   IoT framework components

This case study has been implemented and demonstrated in iCore using the framework introduced in Section 5.3. The following paragraphs describe in a top-down manner how the framework components are used in the implementation of the case study.

Service Request. Starting from the Service Request the customer enters the package with the attached sensor, the starting location and final location of the product. The existing logistic infrastructure calculates the routing and orders the corresponding cooling container and transport vehicles. The iCore infrastructure consults the VO Registry to track the cooling transport.

VO Registry. The package attached sensor registers in the VO Registry. The registration has the information on which gateway the sensor will continuously sent values to an MQTT event bus. Other sensors communicate their values on the same gateway. For example if the gateway is on a truck, the VO Registry has the information, which packages with attached temperature VO are loaded on the truck.

The VO Registry is implemented as an SPRQL-endpoint and keeps track of the relation between sensors and actuators (ICT objects), produce/trucks and other facilities (non-ICT objects), and their virtualization within the iCore Framework (VOs). This allows other components to search for specific non-ICT or ICT objects, for example, based on a specific location. Semantic discovery is used in this use case to establish a relation between trucks and the location of goods. The VO Registry also contains information on how to obtain results from VO functions, for example, how to obtain temperature readings or how to control a HVAC in a warehouse.

VO and CVO Management. The location VO like GPS sensor uses the same gateway. The CVO Management Unit gathers the VO from the VO Management Unit using the same gateway like the attached sensor in the Service Request. The CVO Management Unit initializes several CVO templates based on prediction models and acceptance criteria. After initializing and parameterizing the CVO templates, the CVO queries analyze continuously in real time the values from the VO published on the event bus. If the CVO query detects a possible out-range of the acceptance criteria those values are published on the event bus. Depending on the query other CVO like the smart phone of the truck driver is subscribed on that query and alert the truck driver in real time.

VO Factory. In this use case, the VO Factory requires no user interaction. Since ICT objects push information by themselves, the VO Factory is able to autonomously initiate VOs and register them in the VO Registry. For this purpose, the VO Factory uses the VO Template Repository to resolve (Ambient Systems specific) hardware identifiers to templates for VO Registry entries. In a similar fashion, VO Registry entries are kept up-to-date.

VO Container. The VO Container provisions an execution environment for VOs. The combination VO Back-end/Front-end transforms Ambient Systems specific messages (via socket communication, for example, received from gateways in WSNs) into eXtensible Markup Language (XML)-formatted messages (based on XML templates) and publishes those to an MQTT broker. Device ownership is also stored in the VO Registry. Whenever devices provide information, the ownership of the device is retrieved and VO Front-end and the owner are included in the MQTT topic under which the sensor information is published.

The MQTT topic structure under which messages are published are created following the scheme: *owner/net_id/class/sub_class/function/type/device_id*. The meaning of each field is summarized in the following list:

- *owner*: represents the owner of the specific device. This is derived from a database maintained by the device installer or manufacturer;
- *net_id*: represents the serial identification of the WSN gateway through which the sensor information was received;
- *class/sub_class/function*: identifies the data content, for example, *Temperature* class, *TMP*102 subclass, and *Temperature* function could indicate a temperature reading from a specific sensor type (*Temperature/TMP*102*/Temperature*);
- *type*: indicates the type of the message such as 'Alert'ž tells that the sensor device itself considers the reading to be an alert;
- *device_id*: contains the serial identifier of the sensor device itself. The topic is designed in such fashion that it is easy to create content specific MQTT subscriptions. The VO Registry provides means to compose MQTT topics based on VO function names in addition to the URL of the MQTT broker.

5.5.4   *Information sharing and data confidentiality*

Following an asset lean company approach, it is quite common that transport and logistics companies do not own all facilities required to fulfill their operations. For example, warehouses at airports are often owned by airport facility company instead of the transport company, and so on, while a transport service company needs to act upon data gathered by an ICT infrastructure in the warehouse, which also handles competitive products and deals with competitive transport service providers. The goal of this scenario is to demonstrate that iCore is able by means of policies and PEPs to deal with complex information sharing between stakeholders and keeping critical private information.

The following list summarizes the confidentiality requirements identified in this scenario:

- Device owners may always access (read from/write to) their devices. Others should not be able to access, except in explicit cases.
- Product manufacturers (owners of the perishable product that is routed through the mesh of warehouses) may read devices attached to the product until transfer point to customer. The data has to be anonymized.
- Warehouse operators may read devices that are under their responsibility.
- Transport companies may read devices that are under their responsibility.
- Customers may read upon receipt of the product and acceptance report containing anonymized data.

In this use case, security at VO level is implemented as wrapper around the MQTT broker. The MQTT broker provides publish/subscribe facilities and provides basic access control. This is extended with policy enforcing wrapper that intercepts subscription requests, etc. This enables cross-stakeholder information sharing between users of the VO level, for example, deal with situations where competitor share same warehousing facilities.

5.6   Security engineering for smart business case study

The first step in the application of our methodology is the specification of the structure and behavior model for the smart business case study. Figure 5.6 presents the *MQTT Application* behavior model showing the *Client* and *Broker* interactions used in the implementation of the smart business case study. MQTT clients represent the temperature sensors that publish their temperature values in the MQTT broker running in the WSN gateway device described in the previous section.

The interactions between the Client and Broker behaviors represent the MQTT standard messages, which are intercepted at the broker side by a PEP component. The PEP is a technology-specific extension that generates events to the PDP and receives enforcement actions as a response. The PDP component is accessible through Hypertext Transfer Protocol (HTTP) and is integrated with the SecKit GUI interface that allows the specification of design and runtime models. This enables cross-stakeholder

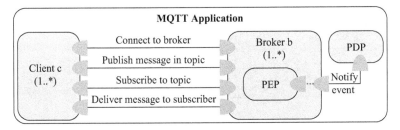

Figure 5.6   MQTT enforcement architecture

information sharing between users of the VO level, for example, to address situations where competitor share same warehousing facilities.

The subscription and signaling of events between the PDP and PEP is done using JavaScript Object Notation (JSON) format and HTTP. In the iCore Framework, we implemented a PEP component [19] embedded in the MQTT broker that is used as a middleware to enable the communication between VOs, CVOs, and services. The PEP component functions as reference monitor and it is implemented using runtime monitoring techniques targeting the specific IoT technology in place. When multiple technologies are used to enable the communication and execution of VOs, CVOs, and services, more PEP components would have to be implemented to monitor and generate events to the PDP component (see Figure 5.1).

Figure 5.7 presents the specification of the same architecture of Figure 5.6 in the SecKit GUI. The PEP behavior is designed as an MQTT authentication plug-in (*MQTT Auth Plugin*), which is the implementation possibility for the Mosquitto MQTT broker with respect to security management extensions. The *IoT System* behavior creates instances of brokers and clients that interact with the respective interaction instantiations defined.

The details of the *Connect to Broker* interaction type show the roles that interact in instantiations of this type (client and broker), and the data and identities exchanged: *clientId* of *MQTT Identity* type and *ipAddress* of *String* type. The data and identity types are specified in the *Data* and *Identity* tabs of the SecKit GUI. The behaviors specified are associated with VO and CVO entity types specified in the *Structure* tab.

After specifying the reference IoT system design, an analysis of threat scenarios can be done in the *Risk* tab. A threat scenario makes reference to the stakeholders, threats, assets, unwanted incidents, vulnerabilities, and countermeasures identified in the IoT system. Figure 5.8 shows one example threat scenario identified in the smart business use case where competitor transport companies may access shipment data of each other. In this threat scenario, a malicious transport company (threat) exploits the lack of access control policies (vulnerability) in order to gain competitive advantage (unwanted incident). This incident causes a leak of confidential information (negative consequence) affecting their competitive advantage (asset). Threat scenarios and negative consequences are marked with quantitative attributes indicating the probability of the scenario (likely) and the impact level (critical), which may be used

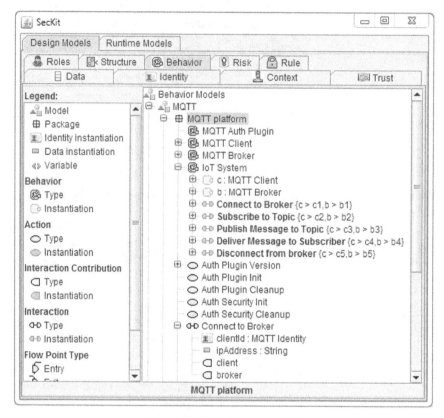

Figure 5.7   Behavior model

for quantitative analysis. In this chapter, we are purely interested in the qualitative specification of threat scenarios without considering quantitative aspects such as the approach proposed by Igor Nai Fovino *et al.* [4].

The example risk model in Figure 5.8 also specifies a countermeasure for the vulnerability of lack of access control, recommending the deployment of a policy rule template where only the responsible for the shipment can read the device information. The rule templates and configurations are specified in the *Rule* tab in the SecKit GUI, presented in Figure 5.9. In this figure, we show the design of the rule template referenced in the identified threat scenario that only allows access to messages published by an MQTT client if the client receiving the message is the one responsible for the shipment. The identifier of the device is retrieved from the MQTT topic information using the variable *deviceId*, which is assigned the result of a regular expression using the topic encoding format introduced in the end of Section 5.5.3. The database associating devices with shipments is accessed in the rule template using a custom action invocation *isResponsibleForShipment*.

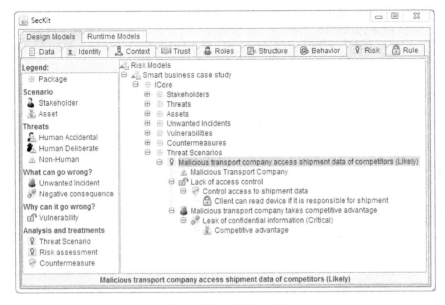

Figure 5.8   Risk model

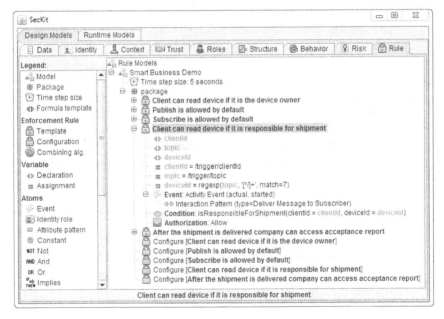

Figure 5.9   Rule model

In this case study, we illustrate the steps in the SecKit methodology and the use of the SecKit toolkit implementation to support the execution of each step in the methodology. The integration of the methodology with this software tool allows for traceability between security engineering requirements and enforceable countermeasures specified using security policy rules. We do not illustrate in this case study the specification of context situations, trust relationships, and more complex security policy rules, we refer the reader to previous publications [20, 21] that describe details about these models and their formalization.

5.7 Discussion

In this chapter, we propose a Model-based methodology and supporting software prototype implementation (SecKit) for security engineering in the IoT domain. We describe the methodology steps and the application of this methodology in a smart business case study proposed by the iCore Project. As a result, all relevant security engineering issues are well documented and traceable back to the IoT system entities. Furthermore, enforceable security policy rules can be specified to enable usage control and protection of user data by mitigating the identified risks.

We have released the SecKit as an open-source project, and this chapter serves as a hands-on tutorial on how to use it in the security engineering process. As an open-source project, the goal is to enable structured community driven specification and sharing of policy templates and technology-specific add-ons focusing on enforcement components for different IoT target technologies and application domains. The adoption of SecKit by many stakeholders has the potential to enable and improve cross-domain security alignment and interoperability.

As future work we plan to focus on the implementation of wizards in our tool to support security engineers in each of the steps of the methodology. We are working on a plug-in mechanism for model extraction and reserve engineering from system runtime artifacts, since it is not straightforward to produce system models for systems already implemented that usually do not have an extensive documentation available. Specifically in the IoT world, our methodology could be integrated with the codebender tool[1] already being used by a large community to design applications for the Arduino platform.

References

[1]   Fenye Bao and Ing-Ray Chen. "Trust management for the Internet of Things and its application to service composition". In *IEEE International Symposium on a World of Wireless, Mobile and Multimedia Networks (WoWMoM)*, pages 1–6, June 2012.

---

[1] See https://codebender.cc/

[2] Patricia Dockhorn Costa, Izon Thomaz Mielke, Isaac Pereira, and João Paulo A. Almeida. "A model-driven approach to situations: Situation modeling and rule-based situation detection". In *EDOC*, pages 154–163. Piscataway, NJ: IEEE, 2012.

[3] Eclipse Foundation. "Eclipse modeling framework project (emf)", 2014. Available at: https://www.eclipse.org/modeling/emf [Accessed 4 April 2016].

[4] Igor Nai Fovino, Marcelo Masera, and Alessio De Cian. "Integrating cyber attacks within fault trees". *Reliability Engineering & System Safety*, 94:1394–1402, 2009.

[5] Sergio Gusmeroli, Salvatore Piccione, and Domenico Rotondi. "A capability-based security approach to manage access control in the Internet of Things". *Mathematical and Computer Modelling*, 58(5):1189–1205, 2013.

[6] Sara Hachem, Thiago Teixeira, and Valérie Issarny. "Ontologies for the Internet of Things". In *Proceedings of the Eighth Middleware Doctoral Symposium, MDS'11*, pages 3:1–3:6, New York, NY: ACM, 2011.

[7] João L. Hernández-Ramos, Antonio J. Jara, Leandro Marín, and Antonio F. Skarmeta. "Distributed capability-based access control for the Internet of Things". *Journal of Internet Services and Information Security (JISIS)*, 3(3/4):1–16, November 2013.

[8] James Irwin. "Digital rights management: The open mobile alliance {DRM} specifications". *Information Security Technical Report*, 9(4):22–31, 2004.

[9] Dongyun Jin. *Making Runtime Monitoring of Parametric Properties Practical*. Ph.D. thesis, University of Illinois at Urbana-Champaign, Champaign, IL, August 2012.

[10] Audun Jøsang. "The right type of trust for distributed systems". In *NSPW'96: Proceedings of the 1996 Workshop on New Security Paradigms*, pages 119–131, Lake Arrowhead, CA, USA, 1996.

[11] Audun Jøsang. "Evidential reasoning with subjective logic". In *13th International Conference on Information Fusion*, Edinburgh, UK, 2010.

[12] SyeLoong Keoh, Kevin Twidle, Nathaniel Pryce, *et al.* "Policy-based management for body-sensor networks". In *Fourth International Workshop on Wearable and Implantable Body Sensor Networks (BSN 2007)*, volume 13 of *IFMBE Proceedings*. Berlin: Springer, 2007.

[13] Zhengqiang Liang and Weisong Shi. "Pet: A personalized trust model with reputation and risk evaluation for p2p resource sharing". In *Proceedings of the 38th Annual Hawaii International Conference on System Sciences (HICSS)*, Hawaii, USA, January 2005.

[14] Mass Soldal Lund, Bjørnar Solhaug, and Ketil Stølen. *Model-Driven Risk Analysis – The CORAS Approach*. Berlin: Springer, 2011.

[15] D. Harrison Mcknight and Norman L. Chervany. "The meanings of trust", 1996. Available at: http://misrc.umn.edu/wpaper/WorkingPapers/9604.pdf [Accessed 4 April 2016].

[16] Patrick O'Neil Meredith, Dongyun Jin, Dennis Griffith, Feng Chen, and Grigore Rosu. "An overview of the mop runtime verification framework".

International Journal of Software Tools for Technological Transfer, 2011. Available at: http://fsl.cs.uiuc.edu [Accessed 4 April 2016].

[17]   Ricardo Neisse and Joerg Doerr. "Model-based specification and refinement of usage control policies". In *11th International Conference on Privacy, Security and Trust (PST)*, Tarragona, Catalonia, Spain, 2013.

[18]   Ricardo Neisse, Igor Nai Fovino, Gianmarco Baldini, Vera Stavroulaki, Panagiotis Vlacheas, and Raffaele Giaffreda. "A model-based security toolkit for the Internet of Things". In *The Ninth International Conference on Availability, Reliability and Security (ARES)*, Fribourg, Switzerland, 2014. Available at: http://ricardo.neisse.name/images/publications/neisse-ares2014.pdf [Accessed 4 April 2016].

[19]   Ricardo Neisse, Gary Steri, and Gianmarco Baldini. Enforcement of security policy rules for the Internet of Things. In *Third International Workshop on Internet of Things Communications and Technologies (IoT-CT), in Conjunction with The 10th IEEE WiMob)*, pages 165–172, Larnaca, Cyprus, October 2014. Available at: http://ricardo.neisse.name/images/publications/neisse-wimob2014.pdf [Accessed 4 April 2016].

[20]   Ricardo Neisse, Gary Steri, Igor Nai Fovino, and Gianmarco Baldini. Seckit: "A model-based security toolkit for the Internet of Things". *Computers & Security*, 54:60–76, October 2015. Available at: http://www.science direct.com/science/article/pii/S0167404815000887 [Accessed 4 April 2016].

[21]   Ricardo Neisse, Maarten Wegdam, and Marten van Sinderen. "Trust management support for context-aware service platforms". In *User-Centric Networking*, Lecture Notes in Social Networks, pages 75–106. Berlin: Springer International Publishing, 2014.

[22]   OMA. "Lightweight machine to machine technical specification". oma ts lightweightm2m id 20130717, draft version 1.0. *OMA*, 2013.

[23]   Simon Pope, Shane Hird, and Matthew Davey. "Subjective logic Java Applet and API", 2014. Available at: http://folk.uio.no/josang/sl/Op.html [Accessed 4 April 2016].

[24]   Dick Quartel. "Action relations – Basic design concepts for behaviour modelling and refinement". Ph.D. thesis, University of Twente, Enschede, The Netherlands, 1998.

[25]   Karl Quinn, David Lewis, Declan O'Sullivan, and Vincent P. Wade. "Trust meta-policies for flexible and dynamic policy based trust management". In *Proceedings of the Seventh IEEE International Workshop on Policies for Distributed Systems and Networks (POLICY)*, pages 145–148, Washington, DC, 2006. Piscataway, NJ: IEEE Computer Society.

[26]   Erik Rissanen. "Extensible access control markup language v3.0", 2010. Available at: http://docs.oasis-open.org [Accessed 4 April 2016].

[27]   A Singh and Ling Liu. "Trustme: Anonymous management of trust relationships in decentralized p2p systems". In *Proceedings of the Third International Conference on Peer-to-Peer Computing (P2P)*, pages 142–149, Linköping, Sweden, September 2003.

[28] ETSI TS. "102 690 machine-to-machine communications m2m functional architecture". *ETSI*, 2011.

[29] ETSI TS. "102 921 machine-to-machine communications (m2m); mia, dia and mid interfaces". *ETSI*, 2012.

[30] C.P. Tso, S.C.M. Yu, H.J. Poh, and P.G. Jolly. "Experimental study on the heat and mass transfer characteristics in a refrigerated truck". *International Journal of Refrigeration*, 25(3):340–350, 2002.

[31] P. Vlacheas, R. Giaffreda, V. Stavroulaki, *et al.* "Enabling smart cities through a cognitive management framework for the Internet of Things". *Communications Magazine, IEEE*, 51(6):102–111, 2013.

[32] W3C. "SPARQL 1.1 overview – w3c recommendation", 2013. Available at: http://www.w3.org/TR/sparql11-overview [Accessed 4 April 2016].

[33] Stephen S. Yau, Yisheng Yao, and Arun Balaji Buduru. "An adaptable distributed trust management framework for large-scale secure service-based systems". *Computing*, 96(10):925–949, 2014.

Chapter 6

Federated Identity and Access Management in IoT systems

Paul Fremantle

Summary

In this chapter, we look at how identity and access management work in the Internet of Things (IoT). In particular, we propose approaches to apply Federated Identity and Access Management (FIAM) techniques to work with IoT devices, networks and servers. We first look at the motivation for using FIAM with IoT. We then look at the previous work and approaches to identity management and access control for IoT systems. We look at how Web-based approaches to FIAM translate into the world of IoT where non-Web protocols are more common. We outline two separate phases of research where we created new FIAM approaches for IoT systems, and we include a performance analysis of the latter system. One emerging area of interest for FIAM is within the domain of Web API Management, and we explore how this also affects the IoT domain. Finally, we conclude with the outcomes of this research, open questions and propose avenues for further research.

6.1 Introduction

The Internet of Things (IoT) refers to the set of devices and systems that intercon-nect real world sensors and actuators to the Internet, which includes many different systems, such as connected cars [1], wearable devices including health and fitness monitoring devices [2], wireless sensor networks [3] and so on. The growth of the number and variety of devices that are collecting data is incredibly rapid. A study by Cisco [4] estimates that the number of Internet-connected devices overtook the human population in 2010, and that there will be 50 billion Internet-connected devices by 2020. Because of the pervasive and personal nature of IoT, privacy and security are important areas for research.

6.1.1 Security and privacy in IoT

There are many security concerns about IoT devices [5, 6]. These include network as well as physical threats to devices. Many devices are based on low-power, inexpensive hardware including 8-bit controllers. These hardware devices are often understood to be unsuitable for high-strength encryption and signature algorithms, although recent research does show that in some cases they can support effective encryption using elliptic curves [7–9].

Limor Fried is a well-known innovator in the IoT field, who has published a "Bill of Rights" for users of IoT devices [10]. One key motivator in this Bill of Rights for this work is the statement that: "Consumers, not companies, own the data collected by Internet of Things devices."

6.1.2 Motivation for Federated Identity and Access Management in IoT

The traditional approach to access control is based on the concept of roles, and is typically managed in a hierarchical, top-down approach [11]. This approach has distinct drawbacks for the IoT. First, it was designed without millions or billions of devices in mind. That would not be an issue, if every device has the same access requirements. However, a fundamental tenet of privacy is that users can decide (and understand) who can share their data. Consumers demand, for example, to allow a specific application access to only certain data. A user might allow their weight-loss club access to a rolling 7-day average but not to individual days weight values. This argues toward a highly controllable model where users can specify authorization for specific devices and/or applications. Cavoukian has promulgated the important concept of *Privacy by Design* and in Reference 12 she states "Users must be empowered to execute effective controls over their personal information."

The second concern with the traditional model is that it utilizes a centralized model of identity and authentication. When there are many devices manufactured by many different organizations and operating in many environments, this is an unrealistic requirement. There are also significant scaling issues with centralized identity models, which are obviated by using a federated model.

A third important concern that is not answered by traditional role-based security models is that of a mechanism for delegation of authority. In many IoT scenarios, there are machines that are operating on behalf of a user, and also scenarios where a device may operate on a third-party's behalf for a specific period of time. For example, the owner of a smart lock may authorize a friend's mobile phone to unlock that door for the next week so that the friend may feed the owner's cat. This argues for a model where the user can *delegate* certain permissions to specific resources to a machine for a limited time.

For this to happen, users must have effective controls over their own data and the ability to specify how this data is shared. This is a major motivation for the use of Federated Identity and Access Management (FIAM) for IoT.

In particular, there are significant challenges around *privacy* with IoT. Many IoT devices are collecting personal data: including human activity, sleep patterns, health information, home automation usage, geographic locations, etc. The result is that access to that data or ability to manipulate those devices may infringe on privacy. As a real example, in 2011, it was publicized that the sexual activity of users of the FitBit activity tracker was public by default [13].

6.1.3 Web API Management

Web Application Programming Interface (APIs) are capabilities offered across the Web that are designed to be accessed by software rather than people. Unlike traditional APIs, Web APIs are inherently public or semi-public in that they are designed to be used over the public Internet and not solely over private networks or Virtual Private Network (VPNs). The public nature of Web APIs poses a number of challenges addressed by the emerging area of *Web API Management* (WAM).

Inevitably, the IoT will need to engage with Web APIs. For now, most IoT devices connect to services that are created by the provider of the hardware, and so are using private APIs. Public APIs are an increasingly important factor. There are a set of companies that are providing common cloud services and corresponding APIs for IoT (such as Xively [14]), and there are emerging API standards for IoT communication (such as HyperCat [15]). Much of the envisioned strength of the IoT will emerge when data from multiple sources can be aggregated, analyzed and acted upon. This will increase the demand for IoT devices to communicate with open Web APIs.

6.1.4 Research questions and contributions

This work addresses some key questions for the use of FIAM and IoT:

- What is the importance of FIAM in the IoT space?
- Can we adapt existing Web-based FIAM technologies to IoT and in particular to lower power binary IoT protocols?
- What are the challenges of using FIAM in the IoT context and what changes need to be made to support its use?
- What are the best architectures for support FIAM with IoT?

In addition, this work addresses the new problem of adapting the principles and technologies of WAM to the landscape of the IoT, which poses challenges stemming from the great numbers and low power of IoT devices, compared to typical full-fledged clients for Web APIs. The problems we are addressing can be clearly stated:

- What is the impact of the IoT onto Web APIs and WAM?
- How do IoT devices identify themselves to Web APIs over IoT protocols?
- How can we add IoT protocol support to existing WAM systems?
- What is the impact of adding identity, usage control and analytics to existing IoT protocol interactions?

This work clearly provides a number of contributions.

- This work provided the first implementation of an FIAM protocol with IoT devices.
- We identified challenges around using existing Web-based protocols (particularly OAuth2) with IoT devices as well as IoT servers.

- We have created prototype systems that implement these approaches, demonstrating that this approach can work in running systems.
- Through the analysis of the first prototype and the development of the second work, we identified significant challenges for using FIAM with IoT, especially with regard to potential hardware attacks on devices, and these were addressed through the inclusion of an additional component (Dynamic Client Registration – DCR). We believe this was the first work to explicitly address this requirement.
- We identified new challenges that emerge from the use of Web APIs from IoT devices, especially those around authentication, usage control and analytics.
- We implemented the first prototype of its kind to add support for IoT specific protocols to API management systems.
- We provide a performance analysis of the IGNITE (Intelligent Gateway for Network IoT Events) system and show that FIAM technologies can be implemented in an effective way for IoT that performs within acceptable ranges.

6.1.5 Outline of the chapter

The remainder of this chapter is structured as follows. In Section 6.2, we review the related work in this space and analyze the gaps in the current literature. In Section 6.3, we look at how we can use OAuth2 and FIAM with IoT systems, including a first prototype – *FIOT* – that provides a framework for analyzing how FIAM and IoT intersect. In Section 6.4, we extend the model of the first prototype into a second, improved model – *IGNITE* – that also explores how FIAM and IoT intersect with the concepts and architecture of WAM. In Section 6.5, we present performance analysis of the IGNITE prototype and analyze the impact of this performance on real world systems. In Section 6.6, we review the work, offer conclusions and suggest areas for further research.

6.2 Related work

In this section, we look at the related work in two areas. First, we examine how FIAM have been applied to IoT. Second, we explore how WAM intersects with IoT and FIAM.

6.2.1 IoT security

The Internet Engineering Task Force (IETF) has published a draft guidance on security considerations for IoT [6]. This draft does discuss both the bootstrapping of identity and the issues of privacy-aware identification. One key aspect is that of bootstrapping a secure conversation between the IoT device and other systems, which includes the challenge of setting-up an encrypted and/or authenticated channel such as those using Transport Layer Security (TLS), Host Identity Protocol (HIP) or Diet HIP. The HIP [16] is a protocol designed to provide a cryptographically secured endpoint to replace the use of Internet Protocol (IP) addresses, which solves a significant problem – IP-address spoofing – in the Internet. Diet HIP [17] is a lighter-weight rendition of the same model designed specifically for IoT and machine-to-machine (M2M) interactions. While HIP and Diet HIP solve difficult problems, they have significant disadvantages to adoption. First, they require low-level changes within the IP stack to implement. Second, as they replace traditional IP addressing they require a major change in many existing systems to work. In addition, neither HIP nor Diet HIP addresses the issues of federated authorization and delegation.

The federated authentication and authorization protocol OAuth [18] (see below Section 6.3) is a relatively new framework and there has been limited research into it as yet. Pai *et al.* [19] have utilized the Alloy Framework [20] to analyze the security constraints of the OAuth protocol.

The MQTT (Message Queue Telemetry Transport) protocol [21] is also a relatively new protocol. Perez [22] has modeled and analyzed the performance of MQTT, and Lee *et al.* [23] have analyzed message loss in MQTT networks and the correlation to the level of Quality of Service (QoS) requested. Mengusoglu and Pickering [24] have created an Autonomic Management system utilizing MQTT as the messaging protocol, whereas Stanford-Clark and Wightwick [25] have demonstrated the applicability of MQTT for environmental monitoring and control systems. Recently, a formal analysis of MQTT's QoS semantics also has been undertaken [26], which demonstrated some ambiguities related to those semantics.

There has been little research into the security of MQTT. The Organization for the Advancement of Structured Information (OASIS) Technical Committee is considering security and in the working draft [27] there is a discussion of how to secure MQTT.

On the other hand, Facebook Messenger [28] on mobile devices uses MQTT and Facebook is also a user of OAuth [29], but there is no published information on whether they utilize the OAuth tokens for MQTT authentication and authorization.

Other related works include the work of Augusto *et al.* [30] have built a secure mobile digital wallet by using OAuth together with the XMPP protocol [31]. While the OAuth protocol has mainly been focused on Web- and Hypertext Transfer Protocol (HTTP)-access control, there is a proposal [32] to support OAuth tokens with the Simple Authentication and Security Layer (SASL) [33] authentication protocol, which demonstrates OAuth usage outside of HTTP.

In Reference 34, we addressed the use of OAuth2 with MQTT, and this is covered in significant detail below in Section 6.3. In IoT-OAS [35], Cirani *et al.* address the use of OAuth2 with Constrained Application Protocol (CoAP), an alternative IoT protocol based on Representational State Transfer (RESTful) principles [36]. It does support the concept of federated identities down to the device, but does not describe the use of the OAuth2 extension – DCR – and the importance of unique OAuth2 credentials per device.

6.2.2 Web API Management

While there is a great deal of industrial effort and research on WAM, the academic literature is sparse. In the industrial sector, much of the literature is provided by vendors. However, the report by Forrester [37] provides a good overview. In the academic literature, Raivio *et al.* [38] explore the business models around Open APIs for the telecommunications industry, and we discuss in Reference 39 the challenges and approaches of managing Web APIs.

In the IoT space, there are a number of efforts around creating open APIs for IoT: for instance, HyperCat [15] is a JavaScript Object Notation (JSON)-based catalogue format for exposing IoT information over the Web, developed by a consortium of academic and industrial partners, and ZettaJS [40] is an open source Web API for IoT devices.

There are a number of existing IoT gateways, including [41–43], that deal with the problem of connecting wireless devices to the wider Internet. They typically bridge multiple low-power devices in a house or factory into a traditional Internet connection. However, our literature search did not identify any server-side gateways/reverse proxies specifically designed for IoT.

We identified two significant gaps in the current literature and existing work in this space. First, most of the work on using APIs with IoT is very limited: there is a common assumption that devices will only communicate with a single API, and there is no discussion of management of these APIs beyond access control. In the access control space, there is a reliance on using outdated models of authentication and authorization (passwords and/or client-side certificates) that are not suitable for device-to-server communication.

Second, when looking at the API Management related work, we found no research that addresses how API Management techniques can be used in the face of IoT specific challenges, especially when using IoT-friendly binary protocols such as MQTT and CoAP. These protocols are important for IoT because of the lower requirements for energy and the lower cost of components required to support them. This encouraged us to explore this area, which is covered in Reference 44, where we addressed the issues regarding how WAM intersects with IoT and we built an enhanced system that demonstrated improved usage of OAuth2 with MQTT. One important contribution of this work is the proposal for using DCR with IoT devices. More information follows in Section 6.4.

6.3   FIAM for IoT

In this section, we outline the work done to create a system called *FIOT* (Federated IoT) which utilizes the OAuth2 protocol to provide authentication and access control for IoT devices and networks that use the MQTT protocol. We first examine the OAuth2 and MQTT protocols, and then we look at how OAuth2 can be applied to MQTT to provide a FIAM solution for IoT.

6.3.1   OAuth

The OAuth 1.0 Protocol [45], and its successor, the OAuth 2.0 Framework [18], are protocols designed to solve the privacy and access control issues related to large-scale Internet-connected applications. OAuth allows users to delegate access to specified functions to third-party Websites. It also allows users to share identification across Websites without sharing their credentials across those Websites. For example, the Twitter Website uses OAuth 2.0 to allow third-party Websites to "tweet" on your behalf.

A more detailed example of the problem which OAuth aims to solve is as follows: social networking Websites often ask users for access to their email contact lists in order to bootstrap or extend the user's social network. In order to implement this, the social network would impersonate the user to their Web-based email provider and access the contact list using the username and password of the user (which was previously asked for). This approach had significant concerns:

- There is no fine-grained access control: once the social network had access to the username and password they could do anything – for example, posting emails or spam.
- There is no time limitation: until the user changes their password the system is now accessible by the third party.
- There is a fundamental difference between M2M interactions, which increasingly uses *tokens* for identification and access, and user-centric security, which uses usernames and passwords.

The OAuth 1.0 Protocol and OAuth 2.0 Authorization Framework are designed to solve these problems. Many Web-based systems nowadays have implemented the OAuth 1.0 and 2.0 protocols, including Google, Facebook, Twitter, GitHub and many others. The protocol is used to protect (and enable) API access and this is clear evidence that it can be used in highly scalable systems.

While there are differences in detail between the two versions we can describe the main approach fairly simply. In order to explore the use of OAuth with IoT, we chose to concentrate on a well-understood protocol for IoT communications, namely MQTT.

There are other approaches and protocols used in the IoT world that would make interesting areas for further research. However, we believe that demonstrating the use of OAuth with MQTT is a useful outcome, and we recommend other areas for research below.

6.3.2 MQTT

The MQTT[1] protocol was originally devised as a protocol for telemetry over constrained networks [46]. Other messaging systems, such as IBM MQ Series [47], assumed high speed networks with low-costs per byte transmission. In many SCADA,[2] IoT and Telemetry examples, the networks are constrained with high-costs per byte (e.g. General Packet Radio Service (GPRS), satellite). Therefore the MQTT specification was designed to have a very low message overhead, with as little as two bytes extra per message.

Another challenge is that IoT devices are often low-powered and reliant on low energy usage. Protocols such as MQTT and CoAP are lower in bandwidth which has a direct effect on energy usage, especially in wireless transmission scenarios. Nicholas [48] shows that MQTT uses considerably less energy that HTTPS in comparative scenarios. This is particularly true in scenarios where notifications need to be sent to devices ("push" scenarios). The traditional way to do this in Web APIs was

---

[1]Originally named the Message Queue Telemetry Transport, but now just MQTT.
[2]Supervisory Control and Data Acquisition.

to require the client to poll the server on a regular basis for updates, which is very expensive in energy and bandwidth usage.

MQTT is used in many IoT scenarios, and there are libraries for microcontroller-based systems such as Arduino [49] that make it easy to utilize.

MQTT is based around a *publish and subscribe* model [50], often known as *pub/sub*. In this model, there are one or more central servers, known as *brokers*, that clients connect to. A client may publish information to the broker, subscribe to receive information or both. The publishers are unaware of the subscribers, and vice-versa. Each publication and subscription is tied to a *topic*. The topic is a named virtual resource that is used to decouple publishers and subscribers. As well as decoupling publishers and subscribers from knowing about each other, the pub/sub model also decouples them in time, because all interactions are asynchronous.

Most pub/sub systems utilize a tree-based model for topics, where there is a hierarchy and wildcard-based matching, which allows subscribers to subscribe to all topics in a branch of the tree, and MQTT also follows this model. For example, a subscriber can subscribe to the topic string "devices/uk/#" which would match topics including "devices/uk/hampshire/emsworth" as well as "devices/uk/sussex." The "#" identifies a wildcard that matches any number of levels within the hierarchy. The "+" character identifies matching only a single level. Publishers cannot publish to a wildcard—they must publish to a fully-qualified topic name.

MQTT has a very simple security model. For authentication, it currently only allows the use of a username and optionally password. For encryption and transport-level security, the TLS standard is recommended, although this is not always suitable for small devices. Some implementations also support the use of a Pre-Shared Key (PSK) with TLS which can be used for authentication as well as encryption. The specification does not describe or recommend any authorization models, but certain implementations support access control lists on specific topics.

We chose to utilize MQTT as the basis of this work for several reasons. MQTT is a de-facto standard protocol used by a variety of IoT and M2M systems. MQTT has been ratified as an International Standard through the OASIS standards body. There are several open source implementations, making it easy to work with. MQTT's central broker model creates an easy place to implement authorization and access control measures. Finally, the topic model is highly flexible and offers scope for authorization control.

6.3.3   FIOT implementation

In order to implement the system, we used several existing open source projects that provided a set of building blocks. This allowed us to focus on the core concerns of authorization and security without spending too much time on implementing aspects that were already understood. There are some areas where the choice of existing technologies provided limitations, and we call out those areas in the results section below.

The overall system consists of four major components: the MQTT broker, the Authorization Server, the Web Authorization Tool (WAT) and the device. The MQTT broker is based on the Mosquitto broker [51], including extensions we created to enable OAuth-based authentication and authorization. The Authorization Server is based on the open source *WSO2 Identity Server* [52]. We made an assumption of a single broker and a single Authorization Server for this prototype. The WAT allows a user or developer to create a token to authorize access to their personal data. In addition, we built and programmed an Arduino-based IoT device which publishes data to the MQTT broker. We provide more details about each of the components below.

Figure 6.1 shows the major components of the system and the major interactions between them.

6.3.3.1 Authorization Server

The WSO2 Identity Server is an open-source identity and access management server that supports a wide variety of identity and access management protocols, including OAuth 2.0 and OpenID Connect amongst many others. The server offers an easy-to-use Web-based console which allows administrators to provision users, configure OAuth applications and other tasks that were needed as part of this exploration.

We did not need to modify or extend the WSO2 Identity Server to implement the "Authorization Server" role defined in our architecture. However, the WSO2 Identity Server implements its own Simple Object Access Protocol (SOAP)-based API for querying the validity and scopes of a token. This was not ideal to call from the message broker, and we preferred the using of the OAuth Token Introspection [53] API, which is a simpler RESTful API. We therefore created [54] a simple bridge between the two APIs using the WSO2 Enterprise Service Bus (ESB) [55], which meant that we could swap the WSO2 Identity Server out for any other OAuth2 server that implements this API. More generally, this extension means any system that uses the Introspection API can now interact with the WSO2 Identity Server.

In order to capture authorization scopes that a token is given access to, we created a JSON [56] encoding. The following example shows a scope that encodes a client who can write to any subtree of */topic/paul* and read/write to */scratch*:

```
[   {rw:"w", topic:"/topic/paul/#"},
    {rw:"rw", topic:"/scratch"} ]
```

This model uses the same syntax for wildcards as the MQTT specification, which is the logical and most effective approach. Because the scope list in OAuth2 is space-delimited, we could not use this JSON as-is to form the authorization scopes. To solve this, we used Base64-encoding to create simple strings that could be used as scopes. There are other solutions we could have taken, but this was expedient as it made parsing the scopes simple for the broker.

6.3.3.2 Web Authorization Tool

We need to be able to instantiate the bearer and refresh tokens and associate the correct access control scopes with these tokens.

This process is described in greater detail in the OAuth 2.0 specification. In the OAuth terms, the MQTT device is the "client." However, we are actually separating this into two parts: the bootstrap process and the runtime process. The bootstrap

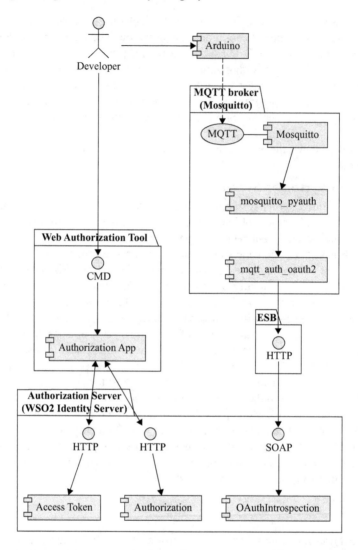

Figure 6.1   Component diagram of the implemented system

process is a set of Web-based flows. Once this process is complete, we then take the bearer and refresh tokens and embed these into our device, which then uses them at runtime. In order to support these Web-based flows, we ideally needed to create a Web-application, because part of the flow is a redirect back to that application, which needs to be Web addressable.

However, we felt a command-line tool would be simpler and more appropriate. To solve this, we created a command-line tool in Python that ran a simple embedded Web-application. This command-line tool spawns a browser with the Uniform Resource Locator (URL) of the Authorization Server to request a token. This is important because the user must authenticate to the Authorization Server, which then redirects that browser back to the local embedded Web server. The result is that there is no need to run a dedicated Web server and the command-line tool can be used to create tokens. This tool generates a textual version of the token, suitable for cutting and pasting into other systems; and generates C code for the Arduino, which can be cut and pasted into the Arduino tooling to update the Arduino. In a real environment as opposed to this prototype, we would have extended the tool to write this directly into the persistent flash memory of a connected device.

In addition, we wanted to demonstrate the other end of the flow – a non-device-based application gaining access to the data published by the IoT device. Of course, this application must be authorized by the owner of the data (the Resource Owner) to be able to access this data. Ideally, this would have been a proper Web application. However, we felt that this angle was the closest to the existing OAuth2 usage and so few insights would be gained from building this, so we emulated this using the default command-line clients provided by Mosquitto. We call this aspect the Message Viewer.

6.3.3.3 The device

To create our sample IoT device, we utilized an 8-bit Arduino open hardware device [49], together with the Arduino Ethernet Shield, which is a daughterboard that provides Ethernet support. The Arduino toolkit has a library that supports the MQTT specification [57]. Our device utilizes a nine-axis inertial measurement unit (IMU) that provides acceleration, rotation and compass data – each in three dimensions, to track the movement of the device. Such devices, when attached to a person, provide significant data on the user's position, activity and exercise levels. Such data exemplifies the problem space here: users wish to share this kind of data, but only in precisely controlled ways, and each user may have radically different approaches and concerns about both data sharing and privacy.

One important point to note is that we did not implement any TLS or encryption from the device to the MQTT broker. This was because of space requirements in the Arduino device. However, we consider this issue orthogonal to the other concerns as OAuth2 and MQTT layer cleanly sits over TLS, without requiring any modification to the flows. Therefore if we had been able to embed a TLS implementation into the system that would not have changed the existing code or the flows.

6.3.3.4 MQTT message broker

Mosquitto is an open-source MQTT broker written in the C language [51]. It is highly portable. Mosquitto has a well-defined model for adding authentication/authorization plugins [58] and there are several third-party plugins but we could not find any example of an existing OAuth-based plugin for Mosquitto.

In this model, the MQTT IoT device connects to the broker, and instead of passing over a username and password, it sends an OAuth 2.0 *bearer token*. The bearer token is a specific type of token supported by the OAuth 2.0 specification that was designed to support very simple client software. Unlike the OAuth 1.0 specification, which required the client to perform signature operations in order to communicate, the bearer token is a token which can be stored and sent as proof of identity. This has its own security challenges which we discuss later.

To create our authorization plugin, we utilized the mosquitto_pyauth [59] project. This plugin converts from the normal C-language based model for Mosquitto plugins into the Python language, allowing us to write the OAuth plugin for Mosquitto in Python instead of C.

We did not assess the performance differences between writing in the native C model versus writing in Python as we were primarily aiming at experimenting with the model and the performance was adequate for our tests. However, it would not be challenging to rewrite the plugin in pure C as most of the work is done by the remote Authorization Server.

Utilizing mosquitto_pyauth as a bridge into Python, we then wrote a Python-based plugin – *mosq-auth-oauth2* – that validates the OAuth2 bearer token which the client passes over. The plugin sends this OAuth token to the *OAuth Introspection Service*, that validates the token and returns a set of authorized scopes for this token. The authentication/authorization plugin then validates if the requested action (reading or writing to a topic) is valid against this scope.

A picture of the implemented architecture is shown in Figure 6.2, where the prototype device is connected to a Mosquitto broker running on Raspberry Pi hardware. Additionally, we include two sequence diagrams showing the system

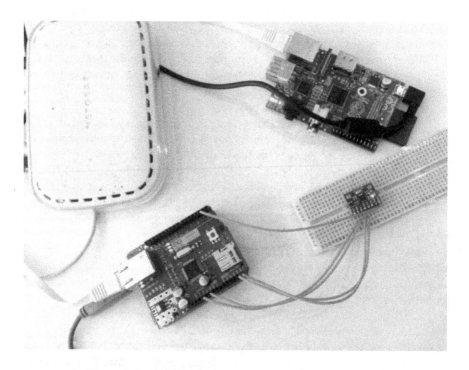

Figure 6.2   Arduino device prototype with nine-axis IMU

interactions: Figure 6.3 shows the sequence diagram of interactions during the bootstrap phase, and Figure 6.4 shows the sequence diagram of the interactions during the use of the device and the Message Viewer.

6.3.3.5   Implementation issues

During the implementation, we came across a number of issues that we considered could be improved in a future implementation. We did not feel that any of these could not be overcome or circumvented, but we did believe that they are worth documenting in order to capture the lessons learnt.

- Because of the nature of the authentication plugin interface to Mosquitto, it was not simple to use the password parameter to pass the token. Therefore, we passed it as the username. This is an implementation specific issue, and does not affect the overall model. However, we would recommend creating a standard way of passing the token over, and we would recommend this be the password field and not the user ID field that was used in this experiment.

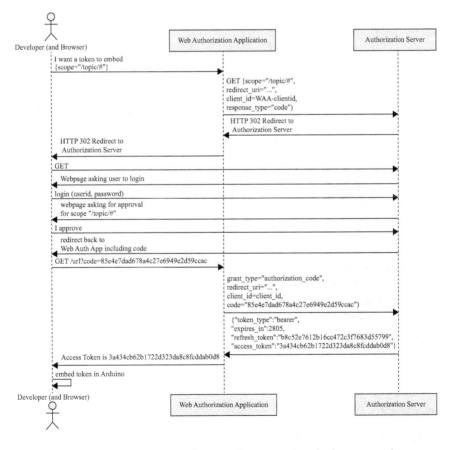

Figure 6.3   UML sequence diagram demonstrating the bootstrap phase

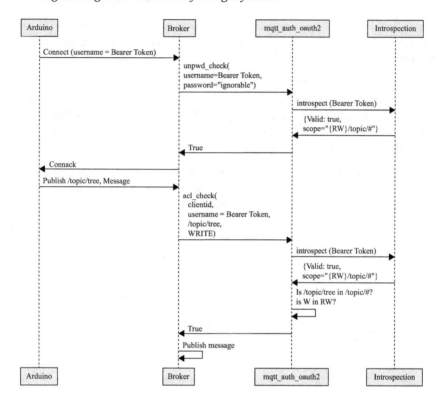

Figure 6.4   UML sequence diagram demonstrating OAuth access control check

- We did not attempt to cache the authorization decisions, or the results of the OAuth introspection. This would be a nice enhancement that would reduce the network traffic and significantly enhance the performance. The fact that the OAuth token has an expiry time gives an effective caching model.
- We did not implement any model where the broker disconnects the client when the token expires, which would be a good extension.
- The Base-64 encoded scopes were not humanly-readable. This meant it was harder for the user to validate the meaning of the scope request. However, in a real consumer facing system, we would expect the scopes to be human-readable names with proper descriptions which would be mapped by the broker into the topic hierarchy.

6.3.4   Results of the FIOT system

Overall the system worked as intended, and showed the following aspects:

- Both IoT and non-IoT clients were able to use OAuth tokens to authenticate to the broker.

- The broker was able to connect using a RESTful interface to the OAuth Authorization Server to validate tokens.
- The broker was able to introspect the token via the RESTful interface on the Authorization Server to retrieve a list of appropriate scopes.
- The broker was able to use the scopes to decide on whether to authorize a publish or subscribe operation.
- The WAT was able to implement the OAuth 2.0 bootstrap process to create Access Tokens, which were then embedded into the MQTT clients.

However, we did find several concerns with the overall model during implementation. We distinguish between minor *implementation issues* which were specific to the chosen implementation technologies and *model issues* which would apply to any attempt to integrate MQTT and OAuth2. We will discuss the model issues next.

6.3.4.1 Refreshing the token

First, the use of HTTP as a protocol to implement the Refresh Token flow is not ideal. It requires the device (in our case the Arduino client) to implement two protocols (MQTT and HTTP), which is inconvenient. To give an example, our IMU and MQTT code took up 97% of the program space of the 8-bit Arduino on which we prototyped. The extra code for OAuth2 took us to 99%. Adding HTTP code for the refresh flow would have pushed us over the limit.

A fundamental issue is that the Refresh Token flow requires the device to produce the Client ID and Client Secret. This is a well-known issue in the mobile space, and highlights a challenge in using the OAuth specification outside the Web application space. This is a limitation in the design of OAuth, which is effectively based around the concept of a Web Application. In the Web Application model, there is a single place where the Web Application is deployed and managed.[3]

In an IoT (or for example the case of a mobile phone application), the client application is deployed in thousands or potentially millions of places. The task of securing the Client Secret is therefore much harder. Microcontrollers such as the Atmel Mega used in the Arduino do allow the code to be locked onto a device, but in any situation where the hardware is accessible, it is always possible for a determined hacker to break such security. This really depends on the value associated with compromising the security: if the reward is high enough, a determined hacker can utilize a Scanning Electron Microscope [60]. The alternative is to issue Access Tokens with very long expiry times, where the expiry time is expected to be longer than the lifetime of the device, or where there is an option to update the device with new firmware. This may be possible, but it goes against best practice for OAuth2.

A key question that arises is whether each device is considered in the OAuth2 world to be a separate Client or that the overall set of devices is considered to be a single Client. If each device is a unique client, then it is to code the Client ID and

---

[3]Of course, a Web Application may be deployed in multiple data centers for scalability and disaster recovery, but this is still logically a single place.

Client Secret into each device. If a single device is compromised, it only compromises that devices' secrets, and they can be revoked. However, it is (as-yet) untested whether OAuth implementations could cope with that many Clients (e.g. millions). In typical Web scenarios there are few Clients (say hundreds or maybe thousands at most).

It seems to us that it is closer to the current model to treat the set of similar devices as *a Client*. In the Web world, a single Website acts as a Client and there are thousands or millions of users, each with a token. The analogy in the IoT world would be a single system with thousands or millions of devices attached. In this model, it would be a serious breach of security to embed the "global" Client ID and Client Secret into each device, as compromising one device would potentially compromise all of them. This concern is more fully explored in Section 6.4.1 were we built an enhanced prototype which supports this.

To investigate this further, we built a second phase to the FIOT experiment, where we mapped the refresh flow into MQTT. Effectively, we created a second MQTT broker – the Authorization Broker (AB) – whose only job is to act as a proxy to the Authorization Server, and we used MQTT instead of HTTP to communicate from the device to this broker.

To explain this, let us name the two brokers: the first (existing) broker is the Data Broker (DB), where data is published and received. The second broker is the AB. The flow works as follows:

- The device attempts to connect to the DB.
- The bearer token has expired and the connection is refused.
- The device connects instead to the AB (with no credentials)
- The device passes the bearer key to the AB
- The AB (or an agent attached to it) adds the Client ID and Client Secret and calls the real Authorization Server to refresh the token.
- The AB then returns the new Bearer Token to the client.
- Finally, the client disconnects from the AB and tries again to connect to the DB with the new Bearer Token.

During the implementation we discovered some interesting sub-issues specific to this refresh flow. First, it is not trivial to send a message to a single device using MQTT. As the system is a pure pub/sub network there is no built-in way of targeting a single device. Obviously, we did not desire to broadcast the refresh token to all devices as that could be misused. To solve this, we implemented a special authentication rule and topic hierarchy on the AB that allowed us to securely send a message just to a single client.

Second, the OAuth specification optionally allows the Authorization Server to refresh the Refresh Token as well. This is a significant issue in a device-oriented world, because the HTTP call to the Authorization Server is unreliable. If the device ends up "out-of-sync" with the Authorization Server, then the device will need to be reregistered. In the Web world, this is easily sorted and happens frequently – it is just a case of the user reauthorizing. With devices, this may not be possible as there is often no Web Browser and UI – which is required in this case. To get round this, we worked with the WSO2 Identity Server development team to add an option to allow us to keep the refresh token constant. We would recommend any OAuth2 server that is being used for IoT devices would avoid changing the Refresh Token with every refresh. This applies to IoT devices using HTTP as well as MQTT or other protocols. Another option would be to use a reliable protocol to transmit the updated refresh token end-to-end from the Authorization Server to the device.

Another concern is whether this model (where any device can connect to the AB and pass over a Refresh Token) is a security hole. Certainly, we could protect it in some regard, but this may not improve matters. If we use a unique credential per device, we have created a circular issue (i.e. how do we manage that credential?). If we use a default credential that is the same for all devices, an attacker needs only to break one device to break this. On the other hand, without a refresh token, the attacker cannot do any real harm. They can cause denial-of-service by issuing many requests, but this could be prevented by standard DoS firewall techniques. On the whole, we felt this model was better than having a unique Client ID/Client Secret per device or having the same Client ID/Client Secret on all devices.

6.3.4.2   Changing the scope

The second issue that was identified during our work was that of updating the scope. In many cases, one might want to change the permissions associated with a device after the firmware has been loaded. In many Web systems, to change the scope of a token is done by reissuing the token. As we have discussed, this is not appropriate for many embedded devices or those without a UI and browser.

A better approach is to update the scope of the existing token. This would remove the burden from the device and deal with it purely at the Authorization Server. However, as currently specified, the OAuth 2.0 specification has no explicit mechanism to allow this. That said, the specification does not prevent it either, but it would need to be an implementation specific approach and therefore would not be interoperable. We see this as an area where the challenges of the IoT should influence the development of OAuth or other similar protocols, as well as the capabilities of OAuth implementations.

6.3.5   *Conclusions of the first phase*

In this section, we have shown that a standardized federated, dynamic, user-directed authentication and authorization model can be adapted from the Web for use in IoT devices. We have argued that this model is important for the concept of privacy with respect to IoT devices and the data that they generate and use. We have identified a number of issues, including both minor implementation issues as well as more fundamental issues where we propose further research. This was the first such implementation of OAuth2 with MQTT. One particularly concerning aspect was the issue around trust in the refresh broker, and the fact that we did not issue each device with its own Access Key and Secret Key. This is a key area that we improved in the next iteration, the IGNITE system (see Section 6.4.1).

6.4   Further exploration of FIAM and IoT, especially with regard to API Management

In this section, we outline an improved approach to applying federated identity and access control to MQTT and IoT devices. We also integrate this into existing WAM systems to provide a wider context for understanding how IoT devices communicate in the overall Web.

Our work addresses the new problem of adapting the principles and technologies of WAM to the landscape of the IoT, which poses challenges stemming from the great numbers and low power of IoT devices, compared to typical full-fledged clients for Web APIs.

Most new Web APIs are working with the OAuth2 [18] standard and utilizing the "Bearer Token" as the API Key. One of the challenges of moving from a model where the API clients are themselves Web servers to a more diverse model where the clients are devices is that the security of these devices is typically much easier to compromise than the security of Web servers. This problem has become apparent with mobile devices. Mobile application developers must embed the OAuth2 credentials into their mobile apps, and because those mobile devices can be "rooted," these credentials can be stolen. There are solutions to this such as Samsung Knox [61], but these are proprietary and only suitable for high-end devices. This rules them out for many IoT devices.

In the FIOT prototype, each device had a refresh token, but did not have their own unique Client Key and Secret Key OAuth credentials. This meant that the refresh broker needed to act as a trusted intermediary on behalf of the device. This diminishes the overall security of the system and exposes situations whereby the access token may be stolen from the device more easily. To address this we envisage that IoT devices will be more likely to need their own OAuth2 credentials per device. It is impractical to think that these client keys will be issued manually to the IoT devices: this process must be automated. This is enabled by the extension to the OAuth2 specification called DCR [62]. The DCR automates the process that a developer would go through on the API portal to gain OAuth2 credentials on behalf of their API client. We built the second IGNITE prototype to explore the use of DCR in IoT scenarios.

We are not aware of any API yet in production where millions of devices each have their own API key, their own set of throttling measures, etc. It can therefore be seen that API management systems will need to evolve to support very large numbers of keys, with millions or even tens of millions of concurrently connected devices.

In summary, this work is addressing how to adapt the existing WAM capabilities to support:

- Large numbers of clients, each with their own credential.
- Devices communicating with public APIs via binary and low-energy protocols such as MQTT and CoAP.
- Usage control, access control, throttling and other API management techniques applied to IoT scenarios.
- FIAM for IoT devices.
- How to apply these capabilities orthogonally to existing systems.

6.4.1 IGNITE – an API gateway for IoT protocols

To solve these issues we prototyped a system that allows using the capabilities of existing API management solutions with IoT protocols. We call the system IGNITE.[4] Our initial work focuses on the MQTT protocol, but in future we intend to extend this to CoAP.

For our proof-of-concept prototype, we extended three major existing open source projects:

- The WSO2 API Manager [63] project provides the main capabilities for WAM including a developer portal, subscription management system, key server, API gateway, access control, throttling, monitoring and analytics system;
- The MITREid-Connect project implements of OAuth2 and OpenID Connect [64] and includes new capabilities such as DCR and Token Introspection.
- The Mosquitto MQTT broker provides an open source messaging broker for the MQTT protocol.

We chose to implement this system in a different model to FIOT for the following reasons.

First, the MITREid-Connect project natively supported a wider set of identity interfaces that the WSO2 Identity Server did at the time of the experiment, including the DCR interface and the Token Introspection interface.

Second, the initial approach of extending the broker had a number of challenges. First, it limited the ways in which we could capture tokens (as documented above) because of the API offered for extending Mosquitto has some inflexibility. More importantly, the gateway approach will be seen to be much more orthogonal and will support working with any broker and not just the Mosquitto system. This is a significant architectural benefit.

To get these benefits, we chose to create an API Management gateway for IoT – a reverse proxy for IoT protocols that plugs into the existing key server architecture and monitoring capabilities.

We currently have built a first prototype of this gateway in Python and we are porting it to Java to improve performance. Figure 6.5 shows the overall architecture with the capabilities of the existing projects plus our added capabilities.

The IGNITE component implements the following logic: On a CONNECT packet arriving, it extracts the OAuth2 Bearer token from the username field in the packet. It then invokes the Token Introspection service on the MITREid-Connect server to validate the token. If the token is valid, the gateway replaces the token in the request with the user ID returned from the introspection call, and forwards the request on to the existing MQTT Broker, which may implement its own validation checks as well. If the token is invalid or no longer active, the IGNITE responds to the client with a packet that indicates that the credential was invalid (a CONNACK packet with ReturnCode=5).

---

[4]The source code is available at https://github.com/pzfreo/ignite.

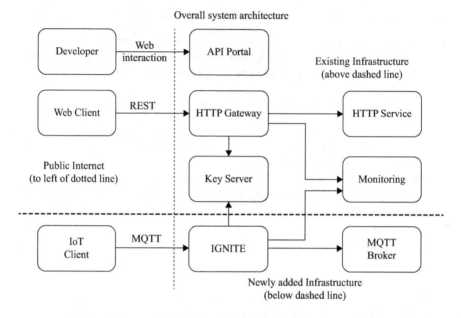

Overall system architecture

Figure 6.5   *System architecture*

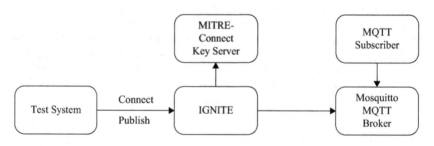

Figure 6.6   *Test architecture*

The monitoring and usage control/throttling aspects of IGNITE are still in development.

6.5   Results

To test the system, we evaluated the performance of this system compared to a direct call to the MQTT broker. In this case, the MQTT broker was not running any authentication of its own, so the comparison is not completely like-for-like. Figure 6.6 shows the architecture of the test set-up.

We used the open source Mosquitto [51] broker as the backend of the tests and ensured that there was a subscriber attached so that the messages would require delivery. For the tests we sampled two flows: A CONNECT flow and a PUBLISH

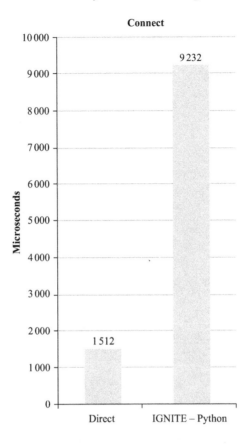

Figure 6.7   CONNECT

flow. For PUBLISH we tested all three levels of QoS: fire-and-forget (QoS0), at least once (QoS1) and exactly-once (QoS2). QoS1 and QoS2 involve multiple packets transferring between the client and the server.

The tests were all run on a single machine[5] using the localhost networking. The gateway tests include both the more functional Python prototype of IGNITE and an early prototype of the Java version. The tests show the average result over 1 000 CONNECT/CONNACK messages and 10 000 PUBLISH messages, in both cases giving the system time to warm up before capturing timing data. The QoS 1 and 2 tests inherently capture the use of PUBACK, PUBREC, PUBREL and PUBCOMP messages. The focus on connection was because the authentication step during connection is where the most work takes place, and on publication because this is the most used flow in MQTT, as subscriptions are rare compared to publication.

The CONNECT results are shown in Figure 6.7. The results show that the overhead of using the Python IGNITE for CONNECT is around 7 700 μs per request. Given

[5]Mac OS/X 10.10 running on a 3 GHz Intel Core i7 with 16 GB RAM and SSD storage.

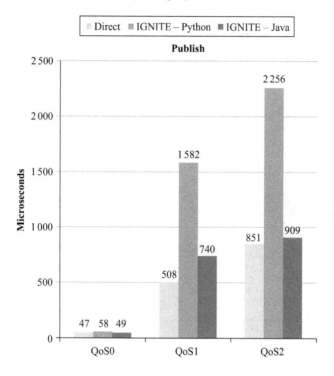

Figure 6.8   PUBLISH

that this includes a HTTP REST call to the key server this is not unexpected. In the WSO2 API Manager, this overhead has been reduced by implementing a binary key validation protocol instead of HTTP. However given that MQTT is a persistent connection compared to existing Web API gateways and HTTP where each request needs to be validated we feel this is a very effective result. We did not (yet) implement caching of token introspection results which could improve this considerably.

The PUBLISH numbers (Figure 6.8) show a much lower overhead. For QoS0 the overhead of going through the IGNITE is around 11 μs. The QoS2 case has a significant higher overhead due to considerably more complex message flow. Even in this case the overhead is less than 1 500 μs and the preliminary data from the Java implementation shows an overhead of less than 60 μs. Note that at this stage we have not yet implemented usage control and monitoring into the PUBLISH flow so these numbers do not yet reflect the full workload required. On the other hand there is as yet no optimization of this prototype code.

To put these numbers into perspective, the typical overhead of such a gateway in the HTTP world is around 500 μs without implementing any OAuth2 token introspection [65]. In addition, these numbers are all likely to be dwarfed by average Internet latencies. For example, the speed of light requires a minimum latency of 40 000 μs between the East and West Coasts of the USA, and typical real-world latencies are twice this. Even with prototype code and no optimization, these numbers are respectable and would fit into the tolerance of many existing IoT projects. Therefore we can conclude that this approach is eminently practicable.

6.6   Discussion

In this work, we have addressed two key aspects. First, we have explored the concepts and realities of FIAM for IoT. Second, we have shown two approaches for implementing this using OAuth2 and MQTT. The second approach offers a significant improvement via the use of DCR.

As a result of this work, there are several recommendations that we propose. These include the need for clear standardization of where to put the token as well as limiting the OAuth2 token size to some reasonable limit – at least for IoT use. Also, there is a need to define a clear MQTT flow for refresh and avoid refreshing the refresh token. Third, there is a requirement to support DCR to ensure that each device has unique credentials. This ensures that a compromise to a single device does not compromise other devices. We also identified a concern with the refresh token being transmitted unreliably, and therefore recommend that refresh tokens are not updated unless there is a reliability mechanism in place.

While there were some issues with implementing FIAM for IoT using OAuth2 with MQTT, the benefits of building on existing widely implemented and deployed protocols are significant. Many years of work and review have gone into the security of OAuth2, and from this work we can state that it is possible to reuse this work with IoT devices and new protocols.

In addition, we found a promising approach to unify IoT devices and MQTT into existing WAM technologies and architecture. We outline our model of enhancing existing WAM systems with a new gateway – IGNITE – that focuses on IoT protocols and demonstrates how protocols such as MQTT can be integrated into existing API Management models with some success, in a completely orthogonal manner. In addition, the model of the server-side IoT gateway that we have introduced with IGNITE offers a considerable number of possibilities for managing usage control, access control, monitoring, etc.

We also present preliminary data on performance which shows that the overheads of such approaches are reasonable even before optimization, caching and other techniques are introduced.

6.6.1   Further research topics

Several areas for further research emerge from this work. A formal model and proof of security attributes for the use of OAuth2 with MQTT and in particular, the refresh flow, would be beneficial. One aspect we did not consider was allowing more than one OAuth2 Authorization Server to be used with a single broker, which would merit research.

In addition, we have identified a set of work on improving the IGNITE gateway: to support CoAP, to optimize performance; to integrate into the monitoring framework; and to support throttling of traffic. In addition we believe there is considerable scope for research to be done on description languages for using MQTT and CoAP for IoT Web APIs.

During discussions of this work with device implementers, a common request has been to create an Arduino "shield" that embodies the MQTT, OAuth2 and TLS behavior completely, which would make an interesting project.

Finally, we believe that some investigation is merited into whether the OpenID Connect specification could add any benefits over OAuth2 in the IoT space.

References

[1]   C. Evans-Pughe, "The connected car," *IEE Review*, vol. 51, no. 1, pp. 42–46, 2005.

[2]   Fitbit, "Fitbit official site for activity trackers and more," 2015 (Visited on 07/09/2015). Available from: http://www.fitbit.com/.

[3]   F. L. Lewis, "Wireless sensor networks," *Smart Environments: Technologies, Protocols, and Applications*, John Wiley & Sons, Inc., Hoboken, NJ, USA, pp. 11–46, 2004.

[4]   D. Evans, "The Internet of Things," *How the Next Evolution of the Internet is Changing Everything, Whitepaper, Cisco Internet Business Solutions Group (IBSG)*, CISCO, 2011.

[5]   R. Roman, P. Najera, and J. Lopez, "Securing the Internet of Things," *Computer*, vol. 44, no. 9, pp. 51–58, 2011.

[6]   e. a. O. Garcia-Morchon, "Security considerations in the IP-based Internet of Things," IETF, Internet Draft, September 2013 (Visited on 04/04/2016). Available from: http://tools.ietf.org/html/draft-garcia-core-security-06.

[7]   M. Sethi, J. Arkko, and A. Keranen, "End-to-end security for sleepy smart object networks," in *Local Computer Networks Workshops (LCN Workshops), 2012 IEEE 37th Conference on*. IEEE, Piscataway, NJ, 2012, pp. 964–972.

[8]   N. Gura, A. Patel, A. Wander, H. Eberle, and S. Shantz, "Comparing elliptic curve cryptography and RSA on 8-bit CPUs," in *Cryptographic Hardware and Embedded Systems – CHES 2004*, series of *Lecture Notes in Computer Science*, M. Joye and J.-J. Quisquater (Eds.), vol. 3156, Springer, Berlin, 2004, pp. 119–132 [Online] (Visited on 04/04/2016). Available from: http://dx.doi.org/10.1007/978-3-540-28632-5_9

[9]   L. Marin, A. Jara, and A. F. Skarmeta, "Shifting primes: optimizing elliptic curve cryptography for smart things," in *Innovative Mobile and Internet Services in Ubiquitous Computing (IMIS), 2012 Sixth International Conference on*. IEEE, Piscataway, NJ, 2012, pp. 793–798.

[10] "A bill of rights for the Internet of Things – room for debate – nytimes.com" (Visited on 11/13/2013) [Online]. Available from: http://www.nytimes.com/roomfordebate/2013/09/08/privacy-and-the-internet-of-things/a-bill-of-rights-for-the-internet-of-things

[11] R. S. Sandhu and P. Samarati, "Access control: principle and practice," *Communications Magazine, IEEE*, vol. 32, no. 9, pp. 40–48, 1994.

[12] A. Cavoukian, "Privacy in the clouds," *Identity in the Information Society*, vol. 1, no. 1, pp. 89–108, 2008.

[13] Zee, "Fitbit users are unwittingly sharing details of their sex lives with the world," 2011 (Visited on 06/04/2013) [Online]. Available from: http://thenext web.com/insider/2011/07/03/fitbit-users-are-inadvertently-sharing-details-of-their-sex-lives-with-the-world/

[14] Xively, "Xively by LogMeIn – business solutions for the Internet of Things" [Online] (Visited on 04/04/2016). Available from: https://xively.com/

[15] R. Lea, IoT Ecosystem Demonstrator Interoperability Working Group. "Hyper-Cat: an IoT interoperability specification," 2013.

[16] R. Moskowitz and P. Nikander, "Host identity protocol architecture," IETF RFC 4423 (2006).

[17] R. Moskowitz and R. Hummen, "Hip diet exchange (dex)," IETF Internet Draft available from https://datatracker.ietf.org/doc/draft-moskowitz-hip-dex/. Last accessed 14 June 2016 (2016).

[18] D. Hardt (Ed.), "The OAuth 2.0 authorization framework," IETF, RFC 6749, October 2012 (Visited on 04/04/2016). Available from: http://www.rfc-editor.org/rfc/rfc6749.txt.

[19] S. Pai, Y. Sharma, S. Kumar, R. M. Pai, and S. Singh, "Formal verification of OAuth 2.0 using alloy framework," in *Communication Systems and Network Technologies (CSNT), 2011 International Conference on*. IEEE, Piscataway, NJ, 2011, pp. 655–659.

[20] D. Jackson, "Alloy 3.0 reference manual," *Software Design Group*, 2004.

[21] D. Locke, "MQ Telemetry Transport (MQTT) V3.1 protocol specification," 2010.

[22] J. Perez, "MQTT performance analysis with OMNeT++," Master's Thesis, Networking Insitut Eurecom; IBM Zurich Research Laboratory Switzerland, September, 2005.

[23] S. Lee, H. Kim, D.-K. Hong, and H. Ju, "Correlation analysis of MQTT loss and delay according to QoS level," in *Information Networking (ICOIN), 2013 International Conference on*. IEEE, Piscataway, NJ, 2013, pp. 714–717.

[24] E. Mengusoglu and B. Pickering, "Automated management and service provisioning model for distributed devices," in *Proceedings of the 2007 Workshop on Automating Service Quality: Held at the International Conference on Automated Software Engineering (ASE)*. ACM, New York, NY, 2007, pp. 38–41.

[25] A. J. Stanford-Clark and G. R. Wightwick, "The application of publish/subscribe messaging to environmental, monitoring, and control systems," *IBM Journal of Research and Development*, vol. 54, no. 4, pp. 1–7, 2010.

[26] B. Aziz, "A formal model and analysis of the MQ Telemetry Transport protocol," in *9th International Conference on Availability, Reliability and Security (ARES 2014)*. IEEE, Piscataway, NJ, 2014.

[27] MQTT version 3.1.1 working draft 15 (Visited on 11/13/2013) [Online]. Available from: https://www.oasis-open.org/committees/download.php/51356/mqtt-v3.1.1-wd15.pdf

[28]   "Building Facebook Messenger" [Online] (Visited on 04/04/2016). Available from: https://www. facebook.com/notes/facebook-engineering/building-facebook-messenger

[29]   "Facebook login" [Online] (Visited on 04/04/2016). Available from: https://developers.facebook. com/docs/facebook-login/

[30]   A. B. Augusto and M. E. Correia, "An XMPP messaging infrastructure for a mobile held security identity wallet of personal and private dynamic identity attributes," in *Proceedings of the XATA*, Vila do Conde, Portugal, 2011.

[31]   P. Saint-Andre, "Extensible messaging and presence protocol (XMPP): Core," 2011.

[32]   K. W. Mills, "A SASL and GSS-API mechanism for OAuth," draft-mills-kitten-sasl-oauth-04.txt, 2011.

[33]   J. Myers, "Simple Authentication and Security Layer (SASL)," IETF RFC 2222, available from https://www.ietf.org/rfc/rfc2222.txt. Last accessed 14 June 2016 (1997).

[34]   P. Fremantle, B. Aziz, P. Scott, and J. Kopecky, "Federated Identity and Access Management for the Internet of Things," in *Third International Workshop on the Secure IoT*, Wroclaw, Poland, 2014.

[35]   S. Cirani, M. Picone, P. Gonizzi, L. Veltri, and G. Ferrari, "IoT-OAS: An OAuth-based authorization service architecture for secure services in IoT scenarios," *Sensors Journal, IEEE*, vol. 15, no. 2, pp. 1224–1234, 2015.

[36]   Z. Shelby, K. Hartke, and C. Bormann, "Constrained Application Protocol (CoAP) draft-ietf-core-coap-18" (Visited on 03/13/2015) [Online]: Available from: https://tools.ietf.org/html/draft-ietf-core-coap-18

[37]   R. Heffner, "The Forrester Wave™: API Management Solutions, Q3 2014," (Visited on 16/06/2016) [Omline]: Available from: (http://unitemybusiness.com/wp-content/uploads/2015/06/450097_forrester_wave_api_management_q3_2014.pdf).

[38]   Y. Raivio, S. Luukkainen, and S. Seppala, "Towards Open Telco-Business models of API management providers," in *System Sciences (HICSS), 2011 44th Hawaii International Conference on*. IEEE, Piscataway, NJ, 2011, pp. 1–11.

[39]   J. Kopecky, P. Fremantle, and R. Boakes, "A history and future of Web APIs," *Information Technology*, 2014, vol. 56, no. 3, pp. 90–97.

[40]   ZettaJS, "Zetta – an API-First Internet of Things (IoT) platform – free and open source software" [Online] (Visited on 04/04/2016). Available from: http://www.zettajs.org/

[41]   Q. Zhu, R. Wang, Q. Chen, Y. Liu, and W. Qin, "IoT gateway: bridging wireless sensor networks into Internet of Things," in *Embedded and Ubiquitous Computing (EUC), 2010 IEEE/IFIP Eighth International Conference on*. IEEE, Piscataway, NJ, 2010, pp. 347–352.

[42]   S. K. Datta, C. Bonnet, and N. Nikaein, "An IoT gateway centric architecture to provide novel M2M services," in *Internet of Things (WF-IoT), 2014 IEEE World Forum on*. IEEE, Piscataway, NJ, 2014, pp. 514–519.

[43]   H. Chen, X. Jia, and H. Li, "A brief introduction to IoT gateway," in *IET International Conference on Communication Technology and Application (ICCTA 2011)*, Beijing, China, 2011, pp. 610–613.

[44]   P. Fremantle, J. Kopecky, and B. Aziz, *Web API Management Meets the Internet of Things*, Springer International Publishing, Switzerland, 2015, Volume 9341, 367–375.

[45]   E. E. Hammer-Lahav, "The OAuth 1.0 protocol," IETF, RFC 5849, April 2010 [Online] (Visited on 04/04/2016). Available from: http://tools.ietf.org/html/rfc5849

[46]   "MQTT history" [Online] (Visited on 04/04/2016). Available from: http://mqtt.org/wiki/doku.php/ history

[47]   L. Gilman and R. Schreiber, *Distributed Computing with IBM MQSeries*. John Wiley & Sons, Inc., New York, NY, 1996.

[48]   S. Nicholas, "Power profiling: HTTPS long polling vs. MQTT with SSL, on Android" (Visited on 06/04/2013). Available from: http://stephendnicholas.com/archives/1217.

[49]   Arduino, "Arduino," 2015 (Visited on 04/04/2016). Available from: http://arduino.cc/

[50]   P. T. Eugster, P. A. Felber, R. Guerraoui, and A.-M. Kermarrec, "The many faces of publish/subscribe," *ACM Computing Surveys (CSUR)*, vol. 35, no. 2, pp. 114–131, 2003.

[51]   Mosquitto, "An open source MQTT v3.1 broker" (Visited on 11/13/2013). Available form: http://mosquitto.org/

[52]   "WSO2 Identity Server – WSO2 Inc." (Visited on 11/13/2013) [Online]. Available from: http://wso2.com/products/identity-server/

[53]   J. Richer, "OAuth 2.0 token introspection," 2013.

[54]   "OAuth2 introspection with WSO2 ESB and WSO2 identity server" [Online] (Visited on 04/04/2016). Available from: http://pzf.fremantle.org/2013/11/oauth2-introspection-with-wso2-esb-and.html

[55]   "Enterprise service bus – WSO2 Inc." (Visited on 11/13/2013) [Online]. Available from: http://wso2.com/products/enterprise-service-bus/

[56]   "The application/JSON media type for JavaScript Object Notation (JSON)," IETF, RFC 4627, July 2006 (Visited on 04/04/2016). Available from: http://www.rfc-editor.org/rfc/ rfc4627.txt

[57]   "Arduino client for MQTT – knolleary" (Visited on 11/13/2013) [Online]. Available from: http://knolleary.net/arduino-client-for-mqtt/

[58]   "jpmens/mosquitto-auth-plug" (Visited on 11/13/2013) [Online]. Available from: https://github.com/jpmens/mosquitto-auth-plug

[59]   "mbachry/mosquitto_pyauth" (Visited on 11/13/2013) [Online]. Available from: https://github.com/mbachry/mosquitto_pyauth

[60]   "Microcontroller unlocking for code recovery – silicon investigations" (Visited on 11/15/2013) [Online]. Available from: http://www.siliconinvestigations.com/Bchip/Code_ext.htm

[61]   Samsung, "Mobile Enterprise Security – Samsung KNOX" (Visited on 03/24/2015). Available from: https://www.samsungknox.com/en, 2015.

[62] N. Sakimura, J. Bradley, and M. Jones, "Final: OpenID Connect Dynamic Client Registration 1.0 incorporating errata set 1" [Online] (Visited on 04/04/2016).   Available   from: http://openid.net/specs/openid-connect-registration-1_0.html

[63] WSO2, "WSO2 API Manager – 100% Open Source API Management Platform – WSO2 Inc." [Online] (Visited on 04/04/2016). Available from: http://wso2.com/products/ api-manager/

[64] J. Richer, D. Greenwood, and B. Bakis, "Componentization of security principles," in *Symposium on Usable Privacy and Security (SOUPS)*, California, USA, 2014.

[65] D. Abeyruwan, "ESB Performance Round 6.5 – WSO2 Inc." (Visited on 03/24/2015). Available from: http://wso2.com/library/articles/2013/01/esb-performance-65/#latency

Chapter 7

On the security of the MQTT protocol

Benjamin Aziz

Summary

We present a security analysis for one of the most popular standards for the Internet of Things, namely the MQ Telemetry Transport (MQTT) protocol, based on a formal model of the protocol in a timed process algebra. We explain the modelling choices we made and give a review of the results of earlier work on the formal verification of the protocol model. We also reveal in this chapter new results related to another case of its message delivery semantics, namely that of at-least-once delivery. We discuss the implications of different failure scenarios for clients and servers in the protocol as well as the variations in the attacker models that could affect the security guarantees of the protocol.

7.1 Introduction

The Internet of Things (IoT) [1] is a new paradigm with the aim of creating connectivity for "everything" that can carry a minimum of storage and computational power, such that these connected things can collaborate anytime, anywhere and in any form, within applications in various personal and social domains such as transportation, enterprise businesses, service and utility monitoring [2, 3]. Some recent estimates suggest that the number of IoT devices exceeds 30 billion with more than 200 billion intermittent connections [4] generating over 700B Euros in revenue by 2020 [5].

This connectivity of IoT devices has been boosted in recent times with the increasing popularity of mobile communications, for example, wireless sensor networks and radio frequency identification technologies, and the proliferation of small hardware with minimum computational and storage capabilities. These technologies coupled with the standardisation efforts of Machine-to-Machine (M2M) communication protocols, such as MQ Telemetry Transport (MQTT) [6] and eXtensible Messaging and Presence Protocol (XMPP) [7], meant that the global vision of the IoT is well within the reach of industries and their markets. However, global applications often require a minimum degree of reliability in terms of the correctness of the specification of the system as well as its level of assurance with respect to nonfunctional properties such as security and privacy. Therefore, there is clear need for adopting formal modelling and analysis methods to ensure that specifications are as little ambiguous as possible leading to more reliable and robust implementations.

Stemming from the above point of view, this chapter adopts a methodology, which can be summarised by the following phases:

- Modelling the IoT protocol (in this case MQTT) in the formal language of TPi.
- Report on the results of applying an abstract static analysis defined in Reference 8 and discuss their security implications.
- Feedback the results of the analysis to the standardisation body (OASIS) responsible for maintaining the MQTT protocol specification.

One should not understate the fact that conducting a formal analysis on a standard specification, such as the one carried out in this chapter, is a time-consuming exercise and requires high expertise levels. However, it can lead to uncovering fundamental issues underlying the informal specifications of standards and hence save even more time later during the system design and implementation phases of the software development life cycle by avoiding errors and weaknesses that could result from such issues.

7.1.1   The MQTT protocol

The MQTT protocol (version 3.1) [6] is described as a lightweight broker-based publish/subscribe messaging protocol that was designed to allow devices with small processing power and storage, such as those which the IoT is composed of, to communicate over low-bandwidth and unreliable networks. MQTT is nowadays used in many business scenarios, for example, the Facebook Messenger [9].

The publish/subscribe message pattern [10], on which MQTT is based, provides for one-to-many message distribution with three varieties of delivery semantics, based on the level of Quality of Service (QoS) expected. The protocol additionally defines the message structure needed in communications between *client*, that is, end-devices responsible for generating data from their domain (the data source) and *servers*, which are the system components responsible for collating source data from clients/end-devices and distributing these data to interested *subscribers*.

In the "at most once" case, messages are delivered with the best effort of the underlying communication infrastructure, which is usually IP-based, therefore there is no guarantee that the message will arrive. This protocol, termed the *QoS = 0 protocol*, is represented by the following flow of messages and actions:

Client → Server   :   **Publish**
  Server Action   :   *Publish message to subscribers*

In the second case of "at least once" semantics, additional mechanisms are incorporated to allow for message duplication, and despite the guarantee of delivering the message, there is no guarantee that duplicates will be suppressed. This case is represented by the following flow of messages and actions:

Client→Server	:	**Publish**
Client Action	:	*Store Message*
Server Actions	:	*Store Message,*
		Publish message to subscribers,
		Delete Message
Server→Client	:	**Puback**
Client Action	:	*Discard Message*

The second message **Puback** represents an acknowledgement of the receipt of the first message, and if **Puback** is lost, then the first message is retransmitted by the client (hence the reason why the message is stored at the client). Once the protocol completes, the client discards the message. This protocol is also known as the $QoS = 1$ *protocol*.

Finally, for the last case of "exactly once" delivery semantics, also known as the $QoS = 2$ *protocol*, the published message is guaranteed to arrive only once at the subscribers. This is represented by the following message/action flow:

Client→Server	:	**Publish**
Client Action	:	*Store Message*
Server Actions	:	*Store Message OR*
		Store Message ID,
		Publish message to subscribers
Server→Client	:	**Pubrec**
Client→Server	:	**Pubrel**
Server Actions	:	*Publish message to subscribers,*
		Delete Message OR
		Delete Message ID
Server→Client	:	**Pubcomp**
Client Action	:	*Discard Message*

Pubrec and **Pubcomp** represent acknowledgement messages from the server, whereas **Pubrel** is an acknowledgement message from the client. The loss of **Pubrec** causes the client to recommence the protocol from its beginning, whereas the loss of **Pubcomp** causes the client to retransmit only the second part of the protocol, which starts at the **Pubrel** message. This additional machinery ensures a single delivery of the published message to the subscribers.

7.1.2   Chapter contributions

An earlier version of this chapter appeared in Reference 8, where a formal analysis of the delivery semantics of the MQTT protocol were presented. The analysis identified the anomalies in the case of QoS $= 2$, which showed that the "exactly once" delivery was not always guaranteed.

The new contributions of the current chapter are:

- For the case of QoS $= 2$, the results presented in Reference 8 led to the updating of the MQTT standard in Reference 11 in relation to the QoS $= 2$ flow of messages in order to remove the ambiguity of its delivery semantics.
- New results regarding the case of QoS $= 1$, where the guarantee that the message is delivered to the subscribers "at least" once is undermined in the presence of a powerful attacker capable of isolating the client devices from the server.
- An informal security discussion, which informs of the various security issues that could arise under attack and failures modes of the protocol.
- An expanded review of related literature.

7.1.3   Chapter structure

The rest of the chapter is organised as follows. In Section 7.2, we describe related work in current literature. In Section 7.3, we provide an overview of the TPi process algebra, a timed version of the π-calculus [12], which we use to model the MQTT protocol. In Section 7.4, we define the MQTT protocol model based on TPi, and explain how this model fits the three delivery semantics that the protocol provides. In Section 7.5, we recap on the analysis of the protocol in the context of its delivery semantics. A version of this analysis was already presented in Reference 8. In Section 7.6, we introduce our notion of a failure as a 0-timed input action, and then discuss informally the different impact this kind of failure could have on the protocol. Finally, in Section 7.7, we informally discuss the security implications of the analysis carried out in the chapter and conclude providing directions for future research.

7.2   Related work

Publish/subscribe is increasingly becoming an important communication paradigm [13], in particular within the domain of sensor device networks and the IoT where messages can be communicated with more efficiency and less consumption of the devices' limited computational power. IBM's MQTT-S protocol [14] was one of the first industrially backed lightweight publish/subscribe protocols that was deployed for wireless sensor and actuator networks. This was followed in year 2010 by version 3.1 [6], which is currently undergoing standardisation by the OASIS community.

There has been very little effort in applying formal analysis tools to IoT communication protocols, mainly due to the novelty of such protocols and their very recent arrival at the scene of communication protocols. On the other hand, some work has been done in the area of publish–subscribe protocols in general. An early attempt in Reference 15 was made to model formally publish/subscribe protocols to capture their essential properties such as minimality and completeness, however, without any attempt to incorporate hostile environments within which these protocols may run.

In Reference 16, the authors define a formal model of publish/subscribe protocols for Grid computing based on Petri-Nets. Their model offers a mechanism for the composition of existing publish/subscribe protocols with model, hence offering a friendly approach for the validation of such protocols. Nonetheless, the focus of their work is mostly on Grid computing scenarios. The work of Reference 17 is an early attempt in discussing security properties and requirements desirable in publish/subscribe protocols, in particular within the domain of Internet-based peer-to-peer systems, where such protocols became popular in their early forms.

Several works, for example, in References 18–22, have adopted model checking as an automated technique for verifying properties of systems to verify properties related to reliability and correctness within the context of publish–subscribe systems, and various levels of efficiency. Garlan *et al.* [18] and Baresi *et al.* [21] define a general framework for model checking publish–subscribe systems, without focusing on specific systems or properties. Their approach is more general than ours, though our approach is more efficient since it handles only one class of pub-sub systems (MQTT) and targets one type of properties. The approach of Reference 19 is closer to our approach in that they adopt a specific system, *thinkteam*, as the target for their analysis. In Reference 20, the authors propose a dedicated model checking technique to verify properties of publish/subscribe-based Message-Oriented Middleware (MOM) systems.

Another close work similar to ours is that of Reference 22, where probabilistic model checking is used to capture uncertainties inherent in publish–subscribe systems. However, despite also being stochastic in nature and hence similar to the TPi language we adopt here, a major difference is that they use probabilities rather than time as a means of expressing that some communications *probably may not* take place. On the same note, probabilistic model checking is also used in Reference 23 to analyse quality of predictions in service-oriented architectures.

Within the domain of sensor network protocols, there is more focus of effort on the formal analysis and verification of such protocols. For example, in Reference 24, the authors apply model checking techniques in the verification of a medium access control protocol called Lightweight Medium Access Control (LMAC). Similarly, in Reference 25, the authors propose a formal model of flooding and gossiping protocols for analysing their performance probabilistic properties. More recently, Heidarian *et al.* [26] proposed a formal model and analysis of clock-synchronised protocols in sensor networks based on timed automata.

7.3 TPi: a timed process algebra

The model of MQTT that we introduce here is based on a process algebra called TPi, originally inspired by Berger and Honda [27] and further developed by Aziz and Hamilton [28]. TPi is a synchronous message-passing calculus capable of expressing timed inputs. The syntax of the language defines processes, $P, Q \in \mathscr{P}$, based on names $x, y \in \mathscr{N}$, as follows:

$$
\begin{array}{llr}
P, Q & ::= & \bar{x}\langle y\rangle.P \mid & \text{output action} \\
& & \texttt{timer}^t(x(y).P, Q) \mid & t\text{-timed input action} \\
& & !P \mid & \text{replication} \\
& & (\nu x)P \mid & \text{new name} \\
& & (P \mid Q) \mid & \text{parallel composition} \\
& & (P + Q) \mid & \text{non-deterministic choice} \\
& & \mathbf{0} \mid & \text{inactive process} \\
& & A(x) & \text{process definition call}
\end{array}
$$

The syntax corresponds to that of the standard synchronous π-calculus [12] except for the fact that input actions are placed within a timer, $\texttt{timer}^t(x(y).P, Q)$, where $t \in \mathbb{N}$ represents a time bound. The input action, $x(y).P$, can synchronise with suitable output actions as long as $t > 0$. Otherwise, when $t = 0$, the timer behaves as Q. There is an assumption that t is decremented by the environment of the process and that t can be any time unit (e.g. tick, second, etc.). Finally, we utilise *process definition calls* in the form of $A(x)$. $A(x)$ calls a process definition $A(y) \overset{\text{def}}{=} P$, and at the same time, passes it the value x to replace y. If the definition A does not accept any input parameters, then we simply omit the input parameter y and write $A() \overset{\text{def}}{=} P$.

In References 8, 28, we defined a nonstandard name-substitution semantics for TPi, which when abstracted using an approximation function α_k, was capable of yielding a finite environment $\phi : \mathscr{N}^\sharp \to \wp(\mathscr{N}^\sharp) \in D_\perp^\sharp$, where $\mathscr{N}^\sharp$ represents the set of abstract names. Unlike $\mathscr{N}$, $\mathscr{N}^\sharp$ is finite and as a result, $\wp(\mathscr{N}^\sharp)$ is also finite. This set is arrived at by means of an abstraction function, α_k, which limits the number of copies of new names created and input parameters in input actions to some finite integer value k. For example, the set of name substitutions $s = \{x_1 \to \{y\}, u_1 \to \{w_1, w_2, w_3, y\}, p_1 \to \{q_1, q_2\}, p_2 \to \{y\}\}$ would be abstracted to the environment $\phi = \{x_1 \to \{y\}, u_1 \to \{w_1, y\}, p_1 \to \{q_1, y\}\}$ for an abstraction value $k = 1$.

The resulting domain of all abstract environments, $D_\perp^\sharp$, was shown in Reference 29 to be an appropriate structure that guarantees termination for an abstract interpretation computed over it (e.g. such as the abstract interpretations defined in References 8, 28, 29). An important element in $D_\perp^\sharp$ is the $\perp_{D^\sharp} = \phi_0$ element, which represents the empty environment such as $\forall x \in \mathscr{N}^\sharp : \phi_0(x) = \{\}$. We also defined in Reference 28 an abstract interpretation based on $D_\perp^\sharp$, which was shown to be safe with respect to the name-substitution semantics, in a similar fashion to previous analyses we defined for different variations of the π-calculus [29–31]. The safety property would guarantee that none of the substitution information is lost, for example, in the case of s above, every substitution is captured by ϕ and the only information lost is the copy number.

7.4 A model of the MQTT protocol

We define a model of the MQTT protocol [6] in TPi as shown in Figure 7.1, which captures the client/server protocol messages. Although the protocol also describes

QoS Level 0 Protocol:

$\quad$ *Client(Publish)* | *Server()*, where:

$\quad$ $Client(z) \stackrel{\text{def}}{=} \overline{c}\langle z \rangle$

$\quad$ $Server() \stackrel{\text{def}}{=} c(x).\overline{pub}\langle x \rangle$

QoS Level 1 Protocol:

$\quad$ *Client(Publish)* | *Server()*, where:

$\quad$ $Client(z) \stackrel{\text{def}}{=} \overline{c}\langle z \rangle.\texttt{timer}^t(c'(y), Client(Publish_{DUP}))$

$\quad$ $Server() \stackrel{\text{def}}{=} !c(x).\overline{pub}\langle x \rangle.\overline{c'}\langle Puback \rangle$

QoS Level 2 Protocol:

$\quad$ *Client(Publish)* | *Server()*, where:

$\quad$ $Client(z) \stackrel{\text{def}}{=} \overline{c}\langle z \rangle.\texttt{timer}^t(c(y).ClientCont(y), Client(Publish_{DUP}))$

$\quad$ $ClientCont(u) \stackrel{\text{def}}{=} \overline{c'}\langle Pubrel_u \rangle.\texttt{timer}^{t'}(c'(w), ClientCont(u))$

$\quad$ $Server() \stackrel{\text{def}}{=} !c(l).(ServerLate(l) + ServerEarly(l))$

$\quad$ $ServerLate(x) \stackrel{\text{def}}{=} (\overline{c}\langle Pubrec_x \rangle.c'(v).\overline{pub}\langle x \rangle.\overline{c'}\langle Pubcomp_v \rangle.$
$\qquad\qquad\qquad\qquad\qquad\qquad\qquad\qquad !(c'(v').\overline{c'}\langle Pubcomp_{v'} \rangle))$

$\quad$ $ServerEarly(x) \stackrel{\text{def}}{=} (\overline{pub}\langle x \rangle.\overline{c}\langle Pubrec_x \rangle.c'(q).\overline{c'}\langle Pubcomp_q \rangle.$
$\qquad\qquad\qquad\qquad\qquad\qquad\qquad\quad !(c'(q').\overline{c'}\langle Pubcomp_{q'} \rangle))$

Figure 7.1 A model of MQTTv3.1 in TPi considering the three levels of QoS

messages between the server and the subscribers, we only focus on one aspect of these, which is the initial publish message from the server to the subscribers. The model expresses three protocols, one for each of the three levels of the QoS specified in Reference 6.

7.4.1 The subscribers

Our model of the subscribers is minimal, since we only care about the first step in their behaviour, which is listening to the published messages announced by the server:

$Subscriber() \stackrel{\text{def}}{=} !pub(x')$

This definition does not care about what happens to the message after it has been read by the subscriber on the channel *pub*. The main reason for including the replication, !, is to allow for the possibility of accepting multiple messages from the server. This will allow us later in the analysis to validate the different delivery semantics associated with the MQTT protocol. The definition can capture the number of times the subscriber will read a message within a single run of the protocol since each instance of x' spawned under the replication is renamed with the labelling system x'_1, x'_2, etc. (following Reference 32). The definition also assumes that the subscriber can wait ad infinitum for a message to be published by the server, and if no such message is published, it will do nothing. This is not realistic, but sufficient for our analysis purposes in this chapter.

7.4.2   The passive attacker

The initial model of the attacker that we consider here is the one who has a primitive (passive) role; that is, one that offers the possibility of consuming the exchanged messages in the protocol only, particularly those messages between the clients and servers. Therefore, its definition is to listen continuously on the channels c and c' over which the client and the server communicate:

$$Attacker() \stackrel{\text{def}}{=} !(c(y') + c'(u'))$$

Similar to the case of the subscribers, the attacker is not in a rush to obtain an input from the protocol, therefore it can wait ad infinitum for a message to be received on its channels c or c' (hence we do not use a timer, or instead it is possible to use a timer with $t = \infty$). It is also possible to define an attacker model with finite input capabilities as follows:

$$Attacker() \stackrel{\text{def}}{=} (c(y_1') + c'(u_1')) \mid \cdots \mid (c(y_n') + c'(u_n'))$$

where the operator ! is replaced by a finite number n of the input choices all composed in parallel. In this finite model, the attacker has a linear behaviour in which it is capable of consuming up to only n messages.

In real terms, the case of a passive attacker represents a lossy network acting as a communication medium for MQTT clients and servers. At this stage, we assume that the attacker is not interested in disrupting the server-subscriber part of the protocol communications. However, we discuss this point informally later in Section 7.7.

7.5   Analysis of the protocol

We now define formally the three message-delivery semantics associated with the MQTT protocol, *at most once*, *at least once* and *exactly once* delivery, and we discuss the results of analysing the protocol in light of these three semantics.

7.5.1   QoS = 0 protocol

The model of QoS = 0 protocol is straightforward. The client process is called from the top level protocol with the *Publish* message. This process is then run in parallel with the server process, which upon receiving the *Publish* message, it publishes it on the *pub* channel where interested subscribers are listening. For simplicity, we assume that the message is published as is. However, a more refined (but not of interest to us) server process would be expected to extract the relevant payload from *Publish* before publishing the actual data. We formalise the semantics of the protocol for QoS = 0 by the following theorem.

Theorem 7.1 (Delivery Semantics for QoS = 0). *The MQTT protocol for the case of QoS = 0 has a delivery semantics of the publish message to the subscribers of "at most once".*

Proof. Given the definition of the subscribers' process in the previous section, a run of this protocol would be equivalent to the following in the absence of any attackers: (*Client*(*Publish*) | *Server*() | *Subscriber*()). Analysing the process renders the following value of ϕ:

$$\phi = \{x \mapsto \{Publish\}, x_1' \mapsto \{Publish\}\}$$

From this, we can see that the message arrives at the subscriber. However, if we re-run the analysis with the attacker process activated: (*Client*(*Publish*) | *Server*() | *Subscriber*() | *Attacker*()), we obtain the following outcome:

$$\phi_{atk} = \{y_1' \mapsto \{Publish\}\}$$

This case shows a run of the protocol, which leads to only y' being instantiated with *Publish*. There is no instantiation of the x or x' variables.

From these results, it is easy to see that there are two possible outcomes. The first value of ϕ represents a normal run where $x' \mapsto \{Publish\}$, whereas in the second value of ϕ_{atk}, we have that $x' \mapsto \{\}$ by the definition of the default state ϕ_0. Hence, it is straightforward to see that the protocol *may* deliver the published message to the subscribers, and therefore, it correctly exhibits the "at most once" delivery semantics. $\square$

For all values of $k > 1$, we do not gain more information about the protocol since the definition of the protocol in Figure 7.1 does not contain replication, which prevents it from being able to interact with an attacker that generates more messages (i.e. a spammy rather than a lossy attacker).

7.5.2 QoS = 1 protocol

The QoS = 1 protocol has a semantics of "at least once" delivery. We model this in Figure 7.1 as a client process, which starts by sending a *Publish* message to the server. The server is capable of inputting this message, publishing it to the subscribers and then replying back to the client with the *Puback* message. Again, for simplicity, we abstract away from the structure of both *Publish* and *Puback*, and point out here that a more refined treatment of these messages (i.e. extracting their payload) does not affect our analysis in the chapter.

The next part is the main difference from the QoS = 0 case above. The client will wait for a finite amount of time, t, on its input channel c' for the *Puback* message from the server. If this message delays (as a result of some communication failure), the client will re-call its process with a new *Publish*$_{DUP}$ message. The difference between *Publish*$_{DUP}$ and *Publish* is that the DUP bit is set in the former to indicate that it is a duplication of the latter. The server on its side is capable of receiving this new publish message since its behaviour is replicated, which means that it can restart its process any number of times required by the context.

The two channels, c and c', distinguish between the two parts of the protocol (i.e. the *Publish* and *Puback* parts). This is not necessary in practice, however it renders our model much simpler by avoiding unnecessary interferences between these two parts. In practice, there would be some message validation mechanisms to prevent such interferences from occurring. We formalise the delivery semantics for this case in terms of the following property.

Property 7.1 (Delivery Semantics for QoS = 1). *The MQTT protocol for the case of QoS = 1 has a delivery semantics of the publish message to the subscribers of "at least once".*   □

In order to understand whether this property holds or not (i.e. whether we can promote it to a theorem or not), we need to analyse both cases when the attacker is present and not. We start first by analysing the protocol under the no-attacker conditions. In this case, we find the following subset value for ϕ and $k = 1$:

$$\phi = \{x_1 \mapsto \{Publish\}, y \mapsto \{Puback\}, x_1' \mapsto \{Publish\}\}$$

This implies normal behaviour, where the published message eventually arrives at the subscriber. Therefore, we can state that for the no-attacker conditions, the protocols holds well for the case of QoS=1. Notably here, the case where $k > 1$ will not add any more information since the server is prompt in replying to the client and so no further calls of the client process definition are made.

Next we run the analysis with the attacker activated, which produces the following subset value of ϕ_{atk}, where $k = 4$ to initiate multiple runs of the protocol:

$$\phi_{atk} = \{x_1 \mapsto \{\}, u_1' \mapsto \{\}, x_1' \mapsto \{\}, y_1' \mapsto \{Publish\}$$
$$x_2 \mapsto \{Publish\}, u_2' \mapsto \{Puback\}, x_2' \mapsto \{Publish\},$$
$$x_3 \mapsto \{Publish_{DUP}\}, u_3' \mapsto \{Puback\}, x_3' \mapsto \{Publish_{DUP}\},$$
$$x_4 \mapsto \{Publish_{DUP}\}, u_4' \mapsto \{Puback\}, x_4' \mapsto \{Publish_{DUP}\}\}$$

Let us consider first the results for the second, third and fourth copies of the captured variables above. We can see from these that the attacker interferes with the protocol by consuming the *Puback* message. In the next two subsets, therefore, the client issues a duplicate $Publish_{DUP}$. In both of these subsets, the attacker continues to consume the acknowledgement message and the client will continue to restart the protocol. Examining these results, we can easily see that the subscribers' input x' has more than one instantiation of the message *Publish*, including when the DUP bit is set. This indicates that the message may arrive more than once at the subscriber. This result is the same for the case of analysis carried out in Reference 8, where only $k = 3$ was considered.

However, let us now examine the first line of results. We notice from this line that the server is not able to receive the message *Publish*. Furthermore, we see that the attacker actually consumes this message instead from the fact that the first copy of y' is instantiated to *Publish*. Analysing this result we can see that neither the server nor the subscriber will be able to obtain the published message under the circumstances where the attacker decides to consume the message directly from the client (different from the above three results, where the attacker does not consume the first message from the client to the server). In a scenario where the attacker repeatedly performs such behaviour (i.e. an isolating attacker), we conclude that the protocol for QoS $= 1$ does not necessarily guarantee a minimum of one delivery of the message to the subscribers, as the protocol specification claims. As the analysis results show, there is likelihood that the message may never be delivered to the subscribers. Hence, Property 7.1 cannot be elevated to a theorem as was the case in QoS $= 0$.

7.5.3 QoS $= 2$ protocol

The last protocol represents the highest level of QoS, indicated by the QoS bit setting of 2. The model of Figure 7.1 contains again the definitions of the client and the publishing server. Similar to (and for the same reasons as) the case of QoS $= 1$, we use two channels for the client: c for the first part ending with the sending of *Pubrec* and c' for the second part ending with the sending of *Pubcomp*.

The client process has two parts. The first could be re-iterated, which will result in the *Publish* message being resent with the DUP bit set in case the *Pubrec* message is not received from the server within a time bound of t units. Note here that the standard protocol of Reference 6 is not clear regarding the resent message. There is no explicit mentioning that the resent publish message is considered different from the original one. The assumption we make is that since DUP is set, then the resent message is a "duplicate" of the original one and therefore it is the same message.

The second part of the client process, *ClientCont*, is instantiated by the first part only if *Pubrec* is received from the server within the time bound t. In this case, it will send a *Pubrel* message to the server parameterised by the same message id as received in the previous message (hence we write *Pubrel$_u$*). After this, it waits for an amount of time t' for the last message from the server, *Pubcomp*, at which point it terminates once this message is received. If this last message does not arrive within the time bound t', it will re-call itself (i.e. the *ClientCont* part), which will result in the re-commencement of the protocol from the point of the sending of the *Pubrel$_u$* message. We believe the above two timed input actions model adequately the requirement "If a failure is detected, or after a defined time period, the protocol flow is retried from the last unacknowledged protocol message; either the PUBLISH or PUBREL." [6, p. 38].

Finally, the last part of the protocol represents the server process. This process after receiving the initial publish message splits into a choice of two processes, *ServerEarly* and *ServerLate*. The main difference between these is whether the publish message is published to the subscribers before or after sending the second message of the protocol *Pubrec$_x$*, which is parameterised by the message id received in the first message from the client.

The specification provides two alternatives for this case [6, p. 38]. The first follows the sequence of actions *store message, publish message and delete message*, whereas the second follows the sequence of actions *store message id, publish message and delete message id*. We term the former a *late publish semantics* and the latter an *early publish semantics*. The specification document states that "The choice of semantic is implementation-specific and does not affect the guarantees of a QoS level 2 flow" [6, p. 38], however, we demonstrate next in terms of the output of our static analysis that this is not generally true.

The whole server process is replicated in order to be able to receive a repeat publish message from the client in the event that $Pubrec_x$ is not received at the client within the time limit. The server process, after sending $Pubrec_x$, goes into the second part of the protocol. In this part, it listens on $c'(v)$ or $c'(q)$ for the incoming $Pubrel$ message from the client. It then continues depending on the choice made earlier to either publish the message and send $Pubcomp_v$ or just send $Pubcomp_q$. In both cases, the $Pubcomp$ message is parameterised by the message id from the received $Pubrel$ message.

The final part now commences, which is a replicated process that again listens for the $Pubrel$ message from the client, and once this is received, it sends another $Pubcomp$ message back to the client. This last part of the server process is similar in both sides of the choice and it will replicate itself until the client receives successfully the $Pubcomp$ message, at which point the client will cease re-sending $Pubrel$ messages.

It is worth noting here that this model assumes that the implementation of the server will cater for a non-deterministic choice of both the early and late publish semantics. However, it is also possible, as we shall see in the next section, to model and analyse the server assuming only one of the two semantics of message publishing is implemented. This would be equivalent to modelling the server process as either $!c(l).ServerLate(l)$ or $!c(l).ServerEarly(l)$.

We now capture the delivery semantics for this protocol in terms of the following property.

Property 7.2 (Delivery Semantics for QoS = 2). *The MQTT protocol for the case of QoS = 2 has a delivery semantics of the publish message to the subscribers of "exactly once".*   □

In the first analysis we run, the attacker is deactivated. We obtain the following subset value for ϕ for $k = 1$:

$$\phi = \{z_1 \mapsto \{Publish\}, l_1 \mapsto \{Publish\}, x_1 \mapsto \{Publish\}, u_1 \mapsto \{Publish\},$$

$$y_1 \mapsto \{Pubrec_x\}, v_1 \mapsto \{Pubrel_u\}, x_1' \mapsto \{Publish, q_1 \mapsto \{Pubrel_u,$$

$$w_1 \mapsto \{Pubcomp_v, Pubcomp_q\}\}$$

The substitutions correspond to normal runs of the protocol for the two choices of the late and early publish semantics, where some values such as for w_1 combine both options. Again, in such normal conditions, increasing k over 1 does not provide any new insight into the protocol delivery semantics.

Next, we examine some of the results of the analysis when the attacker is *activated*. In particular, we consider the case of early publish semantics where we analyse in the context of the server $!c(l).ServerEarly(l)$ (i.e. the server always makes the choice of the $ServerEarly$ process) and the simple attacker model $(c(y') + c'(u'))$. We obtain the following subset of the results, with $k = 2$:

$$\phi_{atk1} = \{x_1 \mapsto \{Publish\}, x'_1 \mapsto \{Publish\}, y'_1 \mapsto \{Pubrec_x\}, x_2 \mapsto \{Publish_{DUP}\},$$
$$x'_2 \mapsto \{Publish_{DUP}\}, y_1 \mapsto \{Pubrec_x\}, q_1 \mapsto \{Pubrel_u\}, w_1 \mapsto \{Pubcomp_q\}, \ldots\}$$

We focus on this interesting subset as it represents a single interference case by the attacker (since $k = 2$). The attacker manages to consume the *Pubrec* message ($y'_1 \mapsto \{Pubrec_x\}$) before the client does so. As a result, the first part of the protocol is repeated and hence, in addition to the initial publish message ($x'_1 \mapsto \{Publish\}$), this leads to a second instance of this message to be announced to the subscribers ($x'_2 \mapsto \{Publish_{DUP}\}$). The second instance then continues as normal, leading to $y_1 \mapsto \{Pubrec_x\}$.

Next, we re-apply the analysis to the case of the full server model and the simple non-replicated attacker model, where again we set $k = 2$ for simplicity:

$$\phi_{atk2} = \{x_1 \mapsto \{Publish\}, x'_1 \mapsto \{Publish\}, y'_1 \mapsto \{Pubrec_x\}, x_2 \mapsto \{Publish_{DUP}\},$$
$$y_1 \mapsto \{Pubrec_x\}, v_1 \mapsto \{Pubrel_u\}, x'_2 \mapsto \{Publish_{DUP}\}, w_1 \mapsto \{Pubcomp_v\}, \ldots\}$$

This subset of the analysis results represents another case of the attacker interfering with the protocol. Unlike the case of the first attack, a different choice of the publish semantics is made here in terms of the re-transmission of first part of the protocol. Here, we find that the *Pubrec_x* acknowledgement message sent by the server is captured by the attacker after an early publish semantics choice is taken involving announcing the publish message to the subscribers (by means of $x'_1 \mapsto \{Publish\}$). This failure in delivering *Pubrec_x* to the client causes a restart of the protocol, however, in this case a different choice is made with the late publish semantics. Continuing with this run, the second part of the protocol causes the duplicated publish message *Publish_{DUP}* to be announced again to the subscribers. Note that this attack would not be possible if either *ServerLate* or *ServerEarly* process only is adopted, but not a choice of both.

These anomalies in the case of QoS $= 2$, particularly those arising from ϕ_{atk2}, were raised as an OASIS issue number MQTT-209 [33] for version. An old paragraph in the specification document [6, p. 38] stated that "The choice of semantic is implementation specific and does not affect the guarantees of a QoS level 2 flow", and as a result of the above results the latest specification of the protocol has been revised to avoid the ambiguity issue [11, p. 55]: "The choice of Method A or Method B is implementation specific. As long as an implementation chooses exactly one of these approaches, this does not affect the guarantees of a QoS 2 flow".

7.6 Client/server timed input failures

So far, we have analysed the MQTT protocol under assumptions that both the client and server do not fail, that is, the loss of messages is entirely a result of communication interferences from the network attacker in its various definitions. Here, we consider the case when both clients and servers fail, and without the presence of the attacker, to analyse the effects of such failures on the protocol.

We use our timed inputs as a mechanism to express failures. More specifically, we define our notion of failure as follows.

Definition 7.1. *A failed input is defined as a 0-timed input,* $\mathtt{timer}^0(x(y).P,Q)$. *This failure will cause the input action to never take place leading to P, and instead the alternative process Q will always be chosen.*

We limit ourselves in this chapter to such clear failures as in the above definition and avoid *semantic failures*, where for an ideal (correct) time value of t, the input $\mathtt{timer}^{t'}(x(y).Q,P)$ is defined such that $t' \neq t$ and $t' > 0$. Such failures require more in-depth analysis of the requirements and design of the system incorporating the MQTT protocol to determine what the right value for t is.

We consider next the impact of input failures on the QoS cases.

7.6.1 The case of QoS = 0

For this case, considering the model of Figure 7.1, the server is the only process that can fail in the above manner since it is the only process that performs an input. We could model this failure in the new server process as follows:

$$Server_{fail} \stackrel{\mathrm{def}}{=} \mathtt{timer}^0(c(x).\overline{pub}\langle x \rangle, \mathbf{0})$$

which results in a failed server, which can never accept the initial message from the client containing the data to be published, *Publish*. This failure can be seen as a denial of service (literally) and would be equivalent to the presence of a passive attacker that repeatedly consumes the client's message without relaying it through to the server. Even in the presence of such failure, the semantics of the protocol in the case of QoS $= 0$ remains intact as specified in Reference 6.

7.6.2 The case of QoS = 1

In the case of QoS $= 1$, there are both possibilities that the client and server processes fail. For the case of the client process, t could be set to 0 in the definition of Figure 7.1 as follows:

$$Client_{fail}(z) \stackrel{\mathrm{def}}{=} \overline{c}\langle z \rangle.\mathtt{timer}^0(c'(y), Client_{fail}(Publish_{DUP}))$$

With such failure in the client process, the client ends up iteratively generating the published message without considering the reply from the server. This amounts to a client with "spammy" behaviour, which will have impact on the server's resources, as the server will repeatedly attempt to reply back to the client's publish messages and possibly keep client connections open, hence draining its resources.

On the other hand, the server itself may incorporate a timed element in its input action that can also fail:

$$Server_{fail}() \stackrel{\mathrm{def}}{=} !\mathtt{timer}^0(c(x).\overline{pub}\langle x \rangle.\overline{c'}\langle Puback \rangle, \mathbf{0})$$

This is equivalent to a server that is experiencing downtime (similar to the case of QoS = 0 above), unable to provide any interactions with the clients.

If we have both the client and server experiencing input-failures, then the effect will be again a spammy client attempting to publish messages to a failed server. The only impact here would be on the communication infrastructure as the useless messages from the client would consume bandwidth needlessly.

7.6.3   The case of QoS = 2

This case introduces a complex failure analysis problem into the discussion. This is due to the fact that the number of possible failed sub-processes in the definition of the protocol for the case of QoS = 2 is 4 (i.e. the processes that contain input actions, *Client, ClientCont, ServerLate* and *ServerEarly*). For the two client processes, *Client* and *ClientCont*, there are four possible definitions:

$$Client_{fail1}(z) \stackrel{\text{def}}{=} \overline{c}\langle z\rangle.\texttt{timer}^0(c(y).ClientCont(y), Client_{fail1}(Publish_{DUP}))$$
$$ClientCont_{fail1}(u) \stackrel{\text{def}}{=} \overline{c'}\langle Pubrel_u\rangle.\texttt{timer}^0(c'(w), ClientCont_{fail1}(u))$$
$$Client_{fail2}(z) \stackrel{\text{def}}{=} \overline{c}\langle z\rangle.\texttt{timer}^0(c(y).ClientCont(y), Client(Publish_{DUP}))$$
$$ClientCont_{fail2}(u) \stackrel{\text{def}}{=} \overline{c'}\langle Pubrel_u\rangle.\texttt{timer}^0(c'(w), ClientCont(u))$$

which are created by setting $t = t' = 0$. The first two processes call internally also failed versions of themselves, whereas the second two definitions call non-failed versions. The call to *ClientCont* from within *Client* has no choices as this call is part of the failed input action meaning it will never take place. On the other hand, the two server subprocesses also have two possible failures for each one of them:

$$ServerLate_{fail1}(x) \stackrel{\text{def}}{=} (\overline{c}\langle Pubrec_x\rangle.\texttt{timer}^0(c'(v).\overline{pub}\langle x\rangle.\overline{c'}\langle Pubcomp_v\rangle.$$
$$!(c'(v').\overline{c'}\langle Pubcomp_{v'}\rangle))))$$
$$ServerLate_{fail2}(x) \stackrel{\text{def}}{=} (\overline{c}\langle Pubrec_x\rangle.c'(v).\overline{pub}\langle x\rangle.\overline{c'}\langle Pubcomp_v\rangle.$$
$$!(\texttt{timer}^0(c'(v').\overline{c'}\langle Pubcomp_{v'}\rangle))))$$
$$ServerEarly_{fail1}(x) \stackrel{\text{def}}{=} (\overline{pub}\langle x\rangle.\overline{c}\langle Pubrec_x\rangle.\texttt{timer}^0(c'(q).\overline{c'}\langle Pubcomp_q\rangle.$$
$$!(c'(q').\overline{c'}\langle Pubcomp_{q'}\rangle))))$$
$$ServerEarly_{fail2}(x) \stackrel{\text{def}}{=} (\overline{pub}\langle x\rangle.\overline{c}\langle Pubrec_x\rangle.c'(q).\overline{c'}\langle Pubcomp_q\rangle.$$
$$!(\texttt{timer}^0(c'(q').\overline{c'}\langle Pubcomp_{q'}\rangle))))$$

These server failure cases represent one of three options for each one, where the last option is redundant and hence not considered:

- *ServerLate* (resp. *ServerEarly*) fails on first input $c'(v)$ (resp. $c'(q)$)
- *ServerLate* (resp. *ServerEarly*) fails on second input $c'(v')$ (resp. $c'(q')$)
- *ServerLate* (resp. *ServerEarly*) fails on both first and second inputs $c'(v)$ and $c'(v')$ (resp. $c'(q)$ and $c'(q')$). This last case is redundant as the failure in the second input is subsumed by the failure in the first input.

Considering all the above failed versions of the client and server processes for the case of QoS = 2, the number of possible combinations of these is $2^8 = 256$, which requires a complex failure analysis.

7.7 Discussion

From the results of the abstract interpretation of all the cases of QoS in the previous sections, we can see that the attacker is always capable of getting hold of the message *Publish* as a result of the attacker being capable of interfering with the communications between the client devices and the servers (brokers). In the case that the published messages are deemed to have some level of secrecy or confidentiality, then the attacker would be compromising this secrecy. In the specification that we considered for the protocol, the channel of communication c is not considered to be private nor encrypted. A private channel would imply some resistance to external interference, for example, by means of some trusted communication hardware module, and the use of encryption as suggested in the MQTT specification can be handled by means of an SSL-based solution, though this would compromise the lightweight nature of the protocol.

In the case of a more active attacker, such as the following one:

$$Attacker() \stackrel{\text{def}}{=} !(c(y') + c'(u')) \mid !(\overline{c}\langle msg\rangle) + \overline{c}'\langle msg\rangle)$$

it would be possible to send any locally defined messages, *msg*, in order to disrupt the flow of the protocol and cause incorrect data to be published to the server. More specifically, if $msg \neq Publish$, then the published data will have no integrity. In fact, it is not difficult to see that if $msg = \_\_$, where $\_\_$ is an empty string, then the QoS = 1 claim of "at least once delivery" is again undermined from message integrity perspective as the message delivered is always an empty string (despite the fact that there is a message being delivered). This is because running the analysis produces the following result:

$$\phi_{atk} = \{x_1 \mapsto \{\}, u'_1 \mapsto \{\}, x'_1 \mapsto \{\}, y'_1 \mapsto \{Publish\}$$
$$x_2 \mapsto \{\_\_\}, u'_2 \mapsto \{Puback\}, x'_2 \mapsto \{Publish\},$$
$$x_3 \mapsto \{\_\_\}, u'_3 \mapsto \{Puback\}, x'_3 \mapsto \{Publish_{DUP}\},$$
$$x_4 \mapsto \{\_\_\}, u'_4 \mapsto \{Puback\}, x'_4 \mapsto \{Publish_{DUP}\}\}$$

This problem can be solved again by means of concealing the communication channel c or digitally signing messages sent over c to preserve their integrity and source authenticity.

Both of the above two issues, that is, secrecy and integrity/authenticity of messages appears in the analysis of Section 7.5 at the client-server end of communications. However, this is by no means the only part where this issue may appear. Consider, for example, a further enhancement of the definition of the attacker:

$$Attacker() \stackrel{\text{def}}{=} !(c(y') + c'(u') + pub(r)) \mid !(\overline{c}\langle msg\rangle) + \overline{c}'\langle msg\rangle + \overline{pub}\langle msg\rangle)$$

In this case, the attacker is also capable of sending and receiving messages on the channel intended for communications between the server and the topic subscribers (i.e. applications) through the use of the *pub* channel. The analysis of the model of MQTT under such an attacker reveals:

$$\phi_{atk} = \{x_1 \mapsto \{\}, u'_1 \mapsto \{\}, x'_1 \mapsto \{\}, y'_1 \mapsto \{Publish\}$$
$$x_2 \mapsto \{Publish\}, u'_2 \mapsto \{Puback\}, x'_2 \mapsto \{\_\},$$
$$x_3 \mapsto \{Publish_{DUP}\}, u'_3 \mapsto \{Puback\}, x'_3 \mapsto \{\_\},$$
$$x_4 \mapsto \{Publish_{DUP}\}, u'_4 \mapsto \{Puback\}, x'_4 \mapsto \{\_\}\}$$

Where the message sent by the attacker over the *pub* channel is assumed to be the null message __. Again this channel could be concealed or rendered secure using cryptography if messages exchanged over it are to be kept out of the attacker's sight.

In the case of an attacker that only exercises input capabilities, example:

$$Attacker() \overset{\text{def}}{=} !(c(y') + c'(u') + pub(r))$$

The most pressing issue will be denial of service by means of message losses. It is easy to see that such an attacker is capable of wiping out any communicated messages:

$$\phi_{atk} = \{x_1 \mapsto \{\}, u'_1 \mapsto \{\}, x'_1 \mapsto \{\}, y'_1 \mapsto \{Publish\}$$
$$x_2 \mapsto \{\}, u'_2 \mapsto \{Puback\}, x'_2 \mapsto \{\},$$
$$x_3 \mapsto \{\}, u'_3 \mapsto \{Puback\}, x'_3 \mapsto \{\},$$
$$x_4 \mapsto \{\}, u'_4 \mapsto \{Puback\}, x'_4 \mapsto \{\}\}$$

We demonstrated how such an issue can arise in the specific case of $QoS = 1$, where a passive attacker with replicated input capabilities will undermine the guarantee that the protocol delivers the published messages "at least once" to the relevant topic subscribers. It is worth noting here that this issue can occur again for all QoS levels on the server-subscriber side of communication.

Future research will be focused on studying the properties of the protocol under more aggressive attacker models and we plan to propose refined versions of the protocol, including the use of lightweight cryptography in scenarios where authentication of the small devices is required. In Reference 34, initial experiments have proven that it is possible to have an authentication/authorisation solution for an MQTT system based on OAuth 2 [35]. In addition, although we carried out a simple modification to the $QoS = 2$ case that removes the duplicated publish message vulnerability, we would like to further investigate in-depth additional mechanisms for improving further the protocol. This would call for more automated approaches, namely using any of a number of automated verification tools that exist in the literature, for example, References 36–41.

References

[1]   J. Gubbi, R. Buyya, S. Marusic, M. Palaniswami, "Internet of Things (IoT): A Vision, Architectural Elements, and Future Directions", *Future Gener. Comput. Syst.* 29 (7) (2013) 1645–1660.

[2]   L. Atzori, A. Iera, G. Morabito, "The Internet of Things: A Survey", *Comput. Netw.* 54 (15) (2010) 2787–2805.

[3]   D. Bandyopadhyay, J. Sen, "Internet of Things: Applications and Challenges in Technology and Standardization", *Wirel. Personal Commun.* 58 (1) (2011) 49–69.

[4]   O. Vermesan, P. Friess, *Internet of Things: Converging Technologies for Smart Environments and Integrated Ecosystems*, The River Publishers, Aalborg, Denmark, 2013.

[5]   O. Mazhelis, H. Warma, S. Leminen, *et al.*, "Internet-of-Things Market, Value Networks, and Business Models: State of the Art Report", Tech. Rep. TR-39 (2013).

[6]   D. Locke, MQ Telemetry Transport (MQTT) V3.1 Protocol Specification. IBM and Eurotech, (2010).

[7]   P. Saint-Andre, K. Smith, R. Tronon, *XMPP: The Definitive Guide Building Real-Time Applications with Jabber Technologies*, O'Reilly Media, Inc., California, USA, 2009.

[8]   B. Aziz, "A Formal Model and Analysis of the MQ Telemetry Transport Protocol", in: *Ninth International Conference on Availability, Reliability and Security (ARES 2014)*, Fribourg, Switzerland, IEEE, Piscataway, NJ, 2014.

[9]   Lucy Zhang, Building Facebook Messenger (Aug. 2011). Available from: https://www.facebook.com/notes/facebook-engineering/building-facebook-messenger/10150259350998920, accessed: 04-04-2016.

[10]   K. Birman, T. Joseph, "Exploiting Virtual Synchrony in Distributed Systems", *SIGOPS Oper. Syst. Rev.* 21 (5) (1987) 123–138.

[11]   A. Banks, R. Gupta, MQ Telemetry Transport (MQTT) V3.1.1 Protocol Specification: Committee Specification Draft 02/Public Review Draft 02 OASIS (2014).

[12]   R. Milner, J. Parrow, D. Walker, "A Calculus of Mobile Processes", *Inf. Comput.* 100 (1) (1992) 1–77.

[13]   A. J. Stanford-Clark, G. R. Wightwick, "The Application of Publish/Subscribe Messaging to Environmental, Monitoring, and Control Systems", *IBM J. Res. Dev.* 54 (4) (2010) 396–402.

[14]   U. Hunkeler, H. L. Truong, A. Stanford-Clark, "MQTT-S – A Publish/Subscribe Protocol for Wireless Sensor Networks", in: *Proceedings of the Third International Conference on COMmunication System softWAre and MiddlewaRE (COMSWARE 2008)*, IEEE, Piscataway, New Jersey, USA, 2008, pp. 791–798.

[15]   R. Baldoni, M. Contenti, S. T. Piergiovanni, A. Virgillito, "Modelling Publish/Subscribe Communication Systems: Towards a Formal Approach", in: *Eighth IEEE International Workshop on Object-Oriented Real-Time Dependable Systems (WORDS 2003)*, IEEE Computer Society, Washington, DC, California, USA, 2003, pp. 304–311.

[16]   L. Abidi, C. Cerin, S. Evangelista, "A Petri-Net Model for the Publish-Subscribe Paradigm and Its Application for the Verification of the BonjourGrid Middleware", in: *Proceedings of the 2011 IEEE International Conference*

on *Services Computing, SCC'11*, IEEE Computer Society, Washington, DC, 2011, pp. 496–503.

[17]   C. Wang, A. Carzaniga, D. Evans, A. Wolf, "Security Issues and Requirements for Internet-Scale Publish-Subscribe Systems", in: *Proceedings of the 35th Annual Hawaii International Conference on System Sciences (HICSS'02)*, Volume 9, IEEE Computer Society, Washington, DC, 2002, pp. 3940–3947.

[18]   D. Garlan, S. Khersonsky, J. S. Kim, "Model Checking Publish-Subscribe Systems", in: *Proceedings of the 10th International Conference on Model Checking Software, SPIN'03*, Springer-Verlag, Berlin, 2003, pp. 166–180.

[19]   M. H. ter Beek, M. Massink, D. Latella, S. Gnesi, A. Forghieri, M. Sebastianis, "Model Checking Publish/Subscribe Notification for Thinkteam®", *Electron. Notes Theor. Comput. Sci.* 133 (2005) 275–294.

[20]   Y. Jia, E. L. Bodanese, C. I. Phillips, J. Bigham, R. Tao, "Improved Reliability of Large Scale Publish/Subscribe Based MOMs Using Model Checking", in: *2014 IEEE Network Operations and Management Symposium, NOMS 2014*, Krakow, Poland, May 5–9, 2014, IEEE, Piscataway, New Jersey, USA, 2014, pp. 1–8.

[21]   L. Baresi, C. Ghezzi, L. Mottola, "On Accurate Automatic Verification of Publish-Subscribe Architectures", in: *Proceedings of the 29th International Conference on Software Engineering, ICSE'07*, IEEE Computer Society, Washington, DC, 2007, pp. 199–208.

[22]   F. He, L. Baresi, C. Ghezzi, P. Spoletini, "Formal Analysis of Publish-Subscribe Systems by Probabilistic Timed Automata", in: *Formal Techniques for Networked and Distributed Systems – FORTE 2007, 27th IFIP WG 6.1 International Conference*, Tallinn, Estonia, June 27–29, 2007, Proceedings, Vol. 4574, Springer, Berlin, 2007, pp. 247–262.

[23]   S. Gallotti, C. Ghezzi, R. Mirandola, G. Tamburrelli, "Quality Prediction of Service Compositions Through Probabilistic Model Checking", in: *Proceedings of the Fourth International Conference on Quality of Software-Architectures: Models and Architectures, QoSA'08*, Springer-Verlag, Berlin, 2008, pp. 119–134.

[24]   A. Fehnker, L. V. Hoesel, A. Mader, "Modelling and Verification of the LMAC Protocol for Wireless Sensor Networks", in: *Proceedings of the Sixth International Conference on Integrated Formal Methods, IFM'07*, Springer-Verlag, Berlin, 2007, pp. 253–272.

[25]   A. Fehnker, P. Gao, "Formal Verification and Simulation for Performance Analysis for Probabilistic Broadcast Protocols", in: *Proceedings of the Fifth International Conference on Ad-Hoc, Mobile, and Wireless Networks, ADHOC-NOW'06*, Springer-Verlag, Berlin, 2006, pp. 128–141.

[26]   F. Heidarian, J. Schmaltz, F. W. Vaandrager, "Analysis of a Clock Synchronization Protocol for Wireless Sensor Networks", *Theor. Comput. Sci.* 413 (1) (2012) 87–105.

[27]   M. Berger, K. Honda, "The Two-Phase Commitment Protocol in an Extended Pi-Calculus", *Electron. Notes Theor. Comp. Sci.* 39 (1).

[28]   B. Aziz and G. Hamilton, "Detecting Man-in-the-Middle Attacks by Precise Timing", in: *Proceedings of the 2009 Third International Conference on Emerging Security Information, Systems and Technologies, SECURWARE'09*, IEEE Computer Society, Washington, DC, 2009, pp. 81–86.

[29]   B. Aziz, G. Hamilton, D. Gray, "A Static Analysis of Cryptographic Processes: The Denotational Approach", *J. Logic Algebr. Program.* 64 (2) (2005) 285–320.

[30]   B. Aziz, "A Static Analysis Framework for Security Properties in Mobile and Cryptographic Systems", Ph.D. thesis, School of Computing, Dublin City University, Dublin, Ireland (2003).

[31]   B. Aziz, G. Hamilton, "The Modelling and Analysis of PKI-Based Systems Using Process Calculi", *Int. J. Found. Comput. Sci.* 18 (3) (2007) 593–618.

[32]   B. Aziz, G. Hamilton, "A Privacy Analysis for the π-Calculus: The Denotational Approach", in: *Proceedings of the Second Workshop on the Specification, Analysis and Validation for Emerging Technologies*, no. 94 in Datalogiske Skrifter, Roskilde University, Copenhagen, Denmark, 2002.

[33]   Paul Fremantle, QOS 2 Delivery Options Must Not Be Mixed (Mar. 2014). Available from: https://issues.oasis-open.org/browse/MQTT-209, accessed: 04-04-2016.

[34]   P. Fremantle, B. Aziz, J. Kopecky, P. Scott, "Federated Identity and Access Management for the Internet of Things", in: *Proceedings of the 2014 International Workshop on Secure Internet of Things (SIoT 2014)*, Wroclaw, Poland, 2014.

[35]   D. Hardt (Ed.), "The OAuth 2.0 Authorization Framework, RFC 6749, IETF". Available from: http://www.rfc-editor.org/rfc/rfc6749.txt, accessed: 04-04-2016.

[36]   "ProVerif: Cryptographic Protocol Verifier in the Formal Model", Available from: http://prosecco.gforge.inria.fr/personal/bblanche/proverif/, accessed: 24-09-2014.

[37]   D. A. Basin, S. Mödersheim, L. Viganò, "OFMC: A Symbolic Model Checker for Security Protocols", *Int. J. Inf. Sec.* 4 (3) (2005) 181–208.

[38]   "The Tamarin Prover for Security Protocol Analysis". Available from: https://hackage.haskell.org/package/tamarin-prover, accessed: 24-09-2014.

[39]   Maude-NPA. Available from:, http://maude.cs.uiuc.edu/tools/Maude-NPA/, accessed: 24-09-2014.

[40]   "Casper: A Compiler for the Analysis of Security Protocols". Available from: http://www.cs.ox.ac.uk/gavin.lowe/Security/Casper/, accessed: 24-09-2014.

[41]   "The Scyther Tool". Available from: http://www.cs.ox.ac.uk/people/cas.cremers/scyther/index.html, accessed: 24-09-2014.

Chapter 8

Securing communications among severely constrained, wireless embedded devices

Alexandros Fragkiadakis, George Oikonomou,
Henrich C. Pöhls, Elias Z. Tragos, and Marcin Wójcik

Summary

The goal of this chapter is to present the ideas and concepts of the EU-FP7 SMART-CITIES project "RERUM" with regards to improving the communication security in Internet of Things (IoT)-based smart city applications. The chapter tries to identify the gaps in previous IoT frameworks with regards to security and privacy and shows the advances that RERUM brings to the IoT community with its significant focus on embedded device functionalities. The goal of the RERUM secure communications framework is to provide light-weight solutions so that they can be applied even in the very constrained IoT devices. Solutions for lightweight encryption (based on the relatively new theory of Compressive Sensing), on transport-layer security (based on DTLS) and on integrity verification of data (using on-device signatures) are presented in detail, discussing their applicability and the benefits they bring to IoT.

8.1 Introduction

The Internet of Things (IoT) is considered to be an important factor of economic growth. This massive network of heterogeneous devices (or interchangeably: machines, things, and objects) is becoming part of peoples' everyday lives and will continue penetrating our day-to-day activities, as smart devices become ubiquitous. This idea of "ubiquitous computing" stems from Mark Weiser, who in 1991 presented several use cases which are still open today and in his time they were considered as "science fiction" [1]. Use cases like the alarm clock that asks the user if she wants coffee and then the coffee maker prepares it or the window that shows the weather forecast for today are only nowadays considered possible with the latest advances of the IoT technologies.

Due to the wide range of possible applications that it can support, the IoT has gained much research attention in the last few years. Everyday objects (e.g. chairs, windows, fridges, books, tables, cars, houses, and trees) are interconnected, communicate with each other, exchange the information they sense, and thus become "smart" equipped with intelligence. However, users and service providers are reluctant to exploit this IoT potential without assurance for the safety of their private information. Why would someone spend money to buy and install an IoT-based home automation system, if his neighbors or any potential burglars would be able to identify when he is at home, what type of devices he has or hack his door and enter the house without any obtrusion? Why would someone use a crowdsourcing application for traffic monitoring if the system would be able to track him and know his location at any given time? Why would someone use smart health products for monitoring his vitals if he is not sure that his health-related information would not be sent to his insurance company to influence the amount of money he pays for his insurance?

With the number of interconnected smart devices increasing exponentially (reports estimate that 50 billion devices will be connected to the Internet by 2020 [2]) new security and privacy issues arise, regarding confidentiality, data integrity, information privacy, and safety. Although there is a lot of work in the literature for handling these issues in standard communication networks and systems, it is not easy to apply existing techniques to IoT systems. The main issue arises from the fact that the IoT does not include only mobile phones, standard laptops, PCs, etc. which are standard communication systems, but it includes mainly devices that are constrained in terms of computational power, memory, and energy, that are standard wireless sensor platforms. Thus, existing sophisticated and standard security/privacy solutions cannot be easily applied on IoT systems with constrained devices.

Security and privacy in the IoT have only recently received a lot of attention, as the technologies become mature to be commercialized. However, the complexity of existing security/privacy solutions for the standard Internet and the resource constraints of the IoT devices are the main reasons for not embedding any type of security/privacy mechanisms on the devices (until recently). Many recent reports indicate a clear lack of security mechanisms for the communication of smart devices that are utilized currently by IoT applications [3–5]. The lack of security mechanisms can result to a number of threats for the users, mostly related with the privacy of their personal information (e.g. health information, name, location, even credit card numbers), the provision of unauthorized access to the systems of the users (even their houses if they are managed by home automation systems) or in extreme situations to affect the health of the users (e.g. if the window actuator malfunctions and closes suddenly hitting the user). From the latter example, it can be seen that security threats in IoT systems can be initiated not only by malicious users, but also from malfunctioning devices.

Thus, in addition to the need to deploy existing sophisticated standard security and privacy solutions in the unconstrained world, the constrained nature of IoT poses itself an additional challenge. This chapter aims to give some ideas and to propose a basic framework for communication security in IoT, focusing on solutions that can be easily implemented on constrained embedded devices.

8.2   Related work

To ensure that an IoT system protects the user's data and the user safety, a number of cross-layer security, privacy, and trust mechanisms have to be taken into account from the design phase [6]. Furthermore, end-to-end security and privacy is needed to ensure strong data protection across all the involved technologies and communication systems that are used to transfer the information from the device to the application [7].

Indeed, due to the fact that IoT systems can involve a number of heterogeneous technologies, the communication layer is one of the main elements that have to be protected from attacks. As it is described in Reference 8, attacks in the communication channel are quite severe because they can result to data loss (either due to interference/jamming or intermediate nodes dropping packets), tampering of data (altered by intermediate malicious hops) or replay attacks. All of these affect the integrity of the data that are used in the IoT applications and can degrade the trustworthiness and the reliability of the overall system.

As explained before, not much work has been done in the areas of communication security and privacy in IoT. This is also proved by the focus of the previous EU-funded IoT projects, which was mainly in the areas of interconnecting systems, analyzing and processing data and exposing services. IoT-A [8] had defined a *Security Functionality Group* that ensures the security and privacy of an IoT system, but with a focus on authentication, authorization, identity and key management and trust. Regarding secure communications, IoT-A focused on a framework for key exchange and management to allow the secure distribution of keys between devices. In i-Core [9], the security-related focus was on enhanced access control using context-awareness and not on communication security or on embedded security on IoT devices. BUTLER [10] had overall a more device-oriented focus compared with other IoT projects, but it's communication security focus was limited on authorization and device authentication. IoT@Work [11] had a cross-layer security component which also included device integrity assurance and network access control, applying also some configuration on network devices. OPENIOT [12] also focused only on service-based authorization, authentication, access control, and identity management. COSMOS [13] is working on hardware embedded security functionalities, however, the proposed mechanisms are not lightweight for standard IoT devices and thus the selected hardware device is a relatively powerful XILINX ZC702 [14]. SMARTIE [15] is also working on authentication, authorization, and on device-oriented topics such as secure storage, but without supporting Datagram Transport Layer Security (DTLS) for securing the Constrained Application Protocol (CoAP) communications.

It is evident from the above, that the main focus of the IoT community was limited on the service-oriented security functionalities for authentication, authorization, and access control. The target of RERUM is to try to secure also the "last mile", namely the communication between the IoT gateways and the constrained devices. The key functionalities presented below are (i) Compressive Sensing (CS)-based encryption, (ii) DTLS-based secure communication, and (iii) integrity protection using on device signatures.

Several contributions have studied CS encryption strength, most of them focusing on the computation secrecy of this scheme. The authors in Reference 16 investigate CS encryption strength for two different types of attacks. For the brute-force case, they show that the complexity of this attack is in the order of $O(N(1/2))$. For the sparsity-related attack, an attacker has to check all possible column combinations of the measurement matrix, something that requires extensive resources, making this attack infeasible. In [17], the authors show that when an attacker decrypts the ciphertext using a wrong measurement matrix, then, the sparsity over the decrypted plaintext is higher than the sparsity of the original one. A multi-class CS encryption scheme is described in Reference 18 where legitimate receivers gain multi-level confidential access based using different measurement matrices.

With the advent of elliptic curve based cryptography (ECC) it was possible to have constrained devices generate signatures, which was not really feasible with RSA based schemes. ECC was presented already in 1986 independently by Miller [19] and Koblitz [20]. Implementation for wireless sensor networks (WSNs) exist, e.g. TinyECC [21], NanoECC [22], or NIST's ECClight [23]; and they have been further optimized [24]. Besides Ayuso *et al.*'s work [24], the 160-bit curve implementation by Kern and Feldhofer [25] or the very lightweight ECC-based construction for authentication of Braun, Hess, and Mayer [26] on radio-frequency identification (RFID)-type devices are all examples showing that ECC is a very good candidate for constrained devices. As the ECC curves previously standardized by NIST have been criticized [27] new curves are en route to standardization, e.g. the signature algorithm named ed25519 [28]. In general, ECC has shown to be usable in software implementations, and is well supported in today's recent IoT hardware platforms [29–31].

8.3   Secure communications for the IoT

In this chapter, the focus is on the communication layer of IoT. We present a framework that can be used for enhancing the communication security and is especially built upon the requirements of networks of constrained devices. We have to note, that in this work, we assume that security includes the concepts of confidentiality and integrity (including authentication and origin) of the information exchanged between constrained IoT devices over a wireless network (of any technology). We start by describing from a forensic examiner's perspective the security issues of the constrained devices and the resulting problems on the IoT applications. Then we present three main communication security techniques as they have been developed in the FP7-SMARTCITIES project RERUM [6, 32]: (i) a lightweight encryption technique using the CS theory [33], (ii) the integrity protection of the data by signing the data to be transmitted from the constrained devices using digital signatures, and (iii) a security mechanism using the DTLS protocol.

The proposed techniques can work either as standalone, separately from each other, or as an integrated communication security framework providing maximum security and protection.

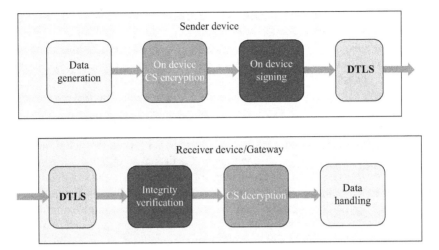

Figure 8.1   Communication security framework

In Figure 8.1, integration of these techniques and the order they can be applied for the transmission of data from one device to another (or the gateway) is depicted. Figure 8.1 is split into layers; the upper part shows the sender device and the lower part shows the receiver. We assume to have a scenario where the IoT device includes one (or more) sensors that gather measurements monitoring the attributes of a physical entity (in the IoT-A terminology [8]), meaning, e.g., the temperature of a room, the status of a window, the energy consumption of the fridge, etc.

Let us assume that in our case, the scenario includes a device with a sensor that monitors the temperature of the living room in a household and is used for managing the air conditioning system. The device transmits periodically the measurements to the gateway. Several attacks on the integrity of the temperature measurements can take place in such scenario, e.g. a malicious user can (i) alter the readings, sending higher temperature values, so that the air conditioning tries to make the room colder or (ii) intercept the temperature values trying to identify the presence of the inhabitant in the room (e.g. when there is a person in the room, the temperature of the room may rise).

The proposed framework can mitigate these attacks by introducing confidentiality and integrity mechanisms on the constrained devices. As depicted in Figure 8.1, the process that is followed for the transmission of the data starts with the encryption of the data using the CS theory. This technique achieves simultaneous encryption and compression of the data, contributing also to the extension of the lifetime of the IoT devices due to the fact that they perform a much smaller number of transmissions, thus consuming less energy. Then, after the encryption, the packets to be transmitted are signed using i.e. classical signatures like Elliptic Curve Digital Signature Algorithm (ECDSA) [34], group signatures [35], or malleable signatures [36]. This will provide authentication and integrity protection, ensuring that any modification of the measurements by unauthorized intermediate nodes will be identified and the modified data will be discarded and not allowed to affect the system's decisions. Finally, the data are being handled by DTLS, which is a cryptographic protocol protecting both the integrity and the confidentiality of the data. Using DTLS again on data that is already encrypted and signed might sound like a repetition, but one shall note that DTLS protects communication channels, and CS and signatures protect the messages regardless if they are in-transit over a communication channel or at rest.

At the receiver side, the opposite order is followed and after DTLS the integrity of the data is verified and if this step is successful then the CS decryption module decrypts and decompresses the data forwarding them to the data handling module (which can be either a service or the routing protocol to forward the data to the IoT middleware, the applications or another device). Of course, the order of these techniques is not fixed and it is left as a design choice of the system administrator, meaning that e.g. the measurements can be first signed one by one and then encrypted and transmitted using DTLS. The order of the application of the techniques does not normally affect the resulting improvement in the security of the framework.

In Figure 8.2, the application of these techniques on the various components of an IoT infrastructure is presented. As it can be seen, the one end of these mechanisms exists on the constrained IoT devices. This means that the encryption of the data using CS, the signing of the data and a part of DTLS are lightweight enough to be running on the constrained devices. Now, the receiving part of DTLS runs mostly on the Gateway (or the destination IoT device in a single hop transmission), although it can also be an end-to-end protocol.

The decryption of the CS-based measurements can be done in various places, according to the design choice of the system administrator. The optimal solution would be to have it on the gateway (which is considered as trusted) so that an adaptive CS-based framework (as the one proposed in Reference 37 can be applied easily without increasing the overall backbone signaling. However, the decryption can also be done in the middleware if the administrator wishes to minimize the complexity of the gateway. In an extreme scenario, the decryption can also be done on the application in order to maximize the protection of the user data that are transmitted (so that no component of the system can identify these data).

Similarly, for the integrity verification, several different options are possible. It can be performed on the gateway, to ensure that all modified data will be discarded at the earliest possible step and improving the reliability of the system, avoiding modified data to be transmitted in the backbone system. Furthermore, the verification can be done on the middleware for reducing the complexity of the gateway. Finally, if the administrator wishes to have end-to-end data integrity protection, the verification can be indeed done also on the applications.

A more thorough analysis of the techniques follows in the next sections of this chapter, discussing in detail the IoT-specific problems they try to solve and how they succeed it being embedded on the constrained IoT devices. However, in order to give an in depth analysis of why it is important to have an embedded security on the devices, the next section provides some results of digital investigations for networks of severely constrained embedded devices, bringing out some of the vulnerabilities of

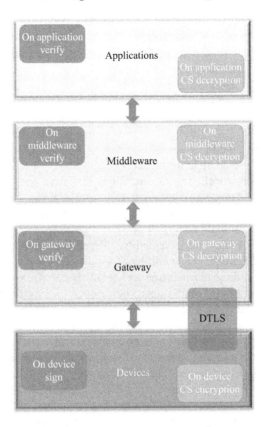

Figure 8.2   Mapping of the communication security techniques on the IoT infrastructure

those networks providing insight on the amount of information that is stored and can be retrieved from devices of this nature by using relatively simple methods and tools.

8.4   Digital investigations for the IoT

Following the discussion in the previous paragraphs and as the number of smart devices keeps increasing, so does the likelihood of malicious activity targeting those devices. Should an attack take place, post-incident forensic analysis can reveal information about the state of the device and network at the time of the attack and ultimately provide clues about the tools used to implement it, or about the attacker's identity. Law Enforcement Agencies (LEAs) currently do not have this capability, which is the topic of ongoing academic research [38].

Here, we focus on TCP/IP-based networks of severely constrained wireless embedded devices. IPv6 over Low Power Wireless Personal Area Networks

(6LoWPANs) [39] and related Internet Engineering Task Force (IETF) specifications [40, 41] have made it possible to use IPv6 in networks of this nature. For those deployments, the IPv6 Routing Protocol for Low-Power and Lossy Networks (RPL) [42] is the de-facto standard for routing.

The adoption of protocols of the TCP/IP family has made those networks vulnerable to new security threats. For instance, the Internet Control Message Protocol Version 6 (ICMPv6) is used to perform key functions in IPv6 networks, one of which is Neighbor Discovery (ND). IPv6 ND's security vulnerabilities have been previously documented [43, 44], while a detailed overview of 6LoWPAN security threats and countermeasures is presented in Reference 45.

Additionally, RPL itself has vulnerabilities that have been discussed by the research and standardization communities: It is susceptible to selective forwarding, wormhole and sinkhole attacks [46–49], which can result in loss of data integrity, availability, and confidentiality. The RPL specification defines some security countermeasures aiming to achieve confidentiality, integrity, and replay protection [42]. Possible solutions have also been contributed by the academic community, such as VeRA [46]. However, these mechanisms are not widely adopted yet. For instance, the RPL implementation in the Contiki open source operating system for the IoT does not support any of them. Among the possible causes are problems with RPL's complexity and implementability, some of which have been previously documented [50]. Additionally, some of RPL's security services rely on the Advanced Encryption Standard (AES). Due to processing constraints of nodes forming 6LoWPANs, there is no well-established method for key negotiation and agreement. Recent research has demonstrated the feasibility of elliptic curve-based approaches [51].

To further clarify the motivation for digital investigation capability for 6LoW-PANs, we use as an example the GINSENG research project. Its outputs demonstrated that it is possible to use a WSN of Contiki-powered nodes to control mission-critical applications in an operational oil refinery [52]. Even though this effort used a bespoke network stack, the control systems industry has been considering a move to open standards, such as TCP/IP and web technologies [53]. Thus, future adoption of 6LoWPAN for industrial automation is not inconceivable. This technological advancement opens up the likelihood of injury or loss of life due to device malfunction or malicious activity. If this were to occur, investigating LEAs should be capable to discover forensic evidence in order to link an attack with the perpetrator.

Conceptually, post-incident analysis of an attack against a 6LoWPAN can be broken down into a number of complementary activities:

1. Analysis of devices forming the network,
2. Analysis of the network's traffic,
3. Analysis of log files

For item 1, we have already published some preliminary results in Reference 38. Briefly, in that work we presented an approach for the extraction and automated analysis of random-access memory (RAM) and flash contents from a constrained wireless sensor node. Based on RAM and flash contents, we demonstrated that it is possible to automatically retrieve network-related information, such as routing table and ND cache contents. Moreover, we demonstrated the capability to correlate information gathered from multiple devices, in order to partially reconstruct the network topology at the time of extraction. The work focused on devices powered by 8051-based, 8-bit micro-controllers running the Contiki open source Operating System for the IoT. However, it also includes an extensive discussion of how it can be extended in order to support different micro-controller architectures and different operating systems.

For item 2, our ongoing work focuses on an automated method for the post-hoc incident detection and analysis of specific 6LoWPAN-specific attacks through an examination of 6LoWPAN traffic captures. Previous work on forensic analysis of WSN traffic has predominantly relied on powerful observer nodes forming part of a WSN deployment. For instance, a network of investigator nodes has been proposed as a solution for digital investigations of wormhole attacks [54]. Those observers are responsible for capturing sensor node behavior and of forwarding this information to the network's base station. The same work proposes a set of algorithms to analyze evidence, in order to identify collaborating malicious nodes and to reconstruct worm-hole attack scenarios. In a similar fashion, it has been demonstrated that a digital forensic readiness layer can be added over a pre-deployed IEEE802.15.4 WSN [55], by the addition of powerful forensic nodes that capture all WSN data plane traffic and maintain frame authenticity and integrity. This work mainly focuses on the reduction of time and cost involved in performing a digital investigation and demonstrates the ability to collect evidence without any modification to an existing network. Powerful observer nodes with the ability to analyze traffic and detect attacks have been adopted by the work documented in Reference 56. Observers can detect various patterns, such as wormhole, black-hole, sinkhole and Sybil attacks. Only illegitimate behavior is forwarded to the network's base station, thus reducing communication overheads. Foren6 is a recent research effort aiming to provide diagnostic and debugging capability for 6LoWPANs [57]. It is a passive monitoring tool capable of collecting information from multiple, potentially mobile sniffers. Foren6 stores a history of network state and topology changes, called versions, and provides the ability to navigate through the entire history in a post-hoc fashion through a network visualizer.

A traffic analysis tool that can identify attacks against RPL and ND in 6LoWPANs, flag events of interest and present results it to an investigator in a human-friendly format can be quite helpful. By combining an analysis of network activity with an analysis of the RAM contents of network nodes, one can reveal useful information about the events that led to a security incident.

Investigations can be further facilitated by a logging infrastructure (item 3). It is possible to log some information on devices themselves, but this has drawbacks: Limited on-device storage capacity means the logs cannot be very detailed. Furthermore, logs will be distributed across the entire deployment and correlation will need to take place. It is also possible to implement a centralized logging infrastructure, for instance by using a syslog server. In this scenario, the drawback is that logging messages would increase network traffic and the approach would not scale well with deployment size and level of required logging detail. Investigations of hybrid approaches, whereby logs are cached locally and sent to a remote location periodically, possibly in an aggregated fashion, have still not been performed and can potentially give very good results.

8.5   CS encryption

Usually, privacy and security become feasible through the use of encryption for the data exchanged between the communicating parties. Several algorithms based on public key encryption involving public and private keys (e.g. RSA [58]), provide robust encryption against unauthorized users. However, this type of algorithms require advanced resources, in terms of processing power and memory; hence, cannot be easily used by the resource-constrained sensors. On the other hand, symmetric algorithms, like the AES [59], require less computational resources and memory, as they use the same key for encryption and compression. The disadvantage of these algorithms is that a key management scheme is required for key distribution to the communicating parties.

Except the privacy and security issues, energy efficiency is also an important issue in WSNs, as sensors are often battery-operated, and they can be placed in harsh environments (e.g. [60]) where human intervention is not possible. As shown in the literature [61, 62], most of the energy spent in a sensor is due to its wireless transceiver operation. Even if the sensor is not active in a wireless communication, it still needs to receive and decode, up to a certain point, all network packets emitted by its neighboring sensors. Usually, data compression is used so as to minimize the required information for transmission, and to save energy. If security and privacy is required, often encryption follows after compression takes place. At the receiver, decryption and decompression are used, as distinct operations, to recover the initial information.

The last few years CS has appeared as a new theory that provides encryption and compression in a single step. As shown in Reference 63, if a signal has a sparse representation in one basis, it can be recovered from a small number of projections in a second basis that is incoherent with the first.

Assume that $x \in \mathbb{R}^N$ refers to information collected by a sensor. Suppose that there is a basis Ψ of $N \times 1$ vectors $\{\psi_{i=1}^N\}$ such that $x = \Psi b$, where $b \in \mathbb{R}^N$ is a sparse vector with S nonzero components ($\|b\|_0 = S$). According to CS theory, the information contained in x can be projected using matrix $\Phi \in \mathbb{R}^{M \times N}$, giving $y = \Phi x$, where $y \in \mathbb{R}^M$ is the compressed version of x. As $M \ll N$, the choice of M controls the compression rate of the original data. Furthermore, the compression rate affects the performance of CS, in terms of the reconstruction error. According to Candes *et al.* [63], an S-sparse signal x can be reconstructed exactly with high probability if $M \geq CS \log(N/S)$, where $C \in R^+$. In any other case, there is a trade-off between the compression rate and the reconstruction error, so, in general, the higher the compression rate is, the higher this error becomes. In general, the compression/encryption using the CS principles is expressed as follows:

$$y = \Phi x = \Phi \Psi b = \Theta b \qquad (8.1)$$

where $\Theta = \Phi \Psi$. The original vector b, and consequently the sparse signal x are estimated using the following ℓ_1 norm convex relaxation problem:

$$\hat{b} = \arg \min \|b\|_1 \quad s.t. \quad y = \Theta b. \qquad (8.2)$$

Observe that the above problem is an under-determined problem with less equations than unknowns as $M \ll N$.

8.5.1 Computational secrecy of CS

Consider a scenario where a wireless sensor repeatedly collects sensitive data x, and then by using Φ encrypts these data into ciphertext y. Also assume that an attacker is present that passively monitors the wireless channel; thus, being able to capture the encrypted data y transmitted by the legitimate sensor. The goal of the attacker is to guess Φ by examining the transmitted blocks of y. This attack is usually referred as *known ciphertext attack*. The attacker may try to guess Φ by forcing a brute force attack based on the ciphertexts it has collected, and searching over the values for Φ based on a step size. Then, and for each ciphertext, it creates (guesses) a matrix Φ', and it reconstructs (decrypts) y to $\hat{x} = \Psi\hat{b}$ using (8.2). At this point, the attacker can estimate, through the reconstruction process (see Reference 37 for details), the residual error that can be used as a metric of the reconstruction accuracy. If the residual error is larger than a threshold, it retries the same procedure, otherwise, it stops and assumes that it has guessed the correct matrix Φ. Nevertheless, as shown in Reference 16, for this brute force attack to become feasible, the computation cost is in the order of $O(N^{1.2})$, something that makes this process too expensive.

Another type of attack against a CS crypto-system is an attack based on the symmetry and sparsity structure of matrix Φ (described in Reference 16). This attack is composed of two phases: During the first phase, the attacker tries to estimate the t leading columns of matrix Φ, assuming that x has t nonzero leading coefficients, and the corresponding coefficients in x such that $\Phi_t x_t = y$. A random permutation of the columns of Φ, and of the corresponding positions in x, produces the same values of y. For this reason, during the second phase, the attacker has to determine the appropriate permutation, so as to find a suitable solution for the over-determined system shown in (8.2). This system has become over-determined as $t < N$. The number of possible permutations requires $C(N, t) \times t!$ possible arrangements that make this attack highly complex.

8.5.2 Information theoretic secrecy of CS

Information theoretic secrecy is based on the statistical properties of a crypto-system providing security even if an attacker has an unbounded processing power. Shannon [64] introduced the idea of perfect secrecy, defining that a crypto-system that achieves perfect secrecy if the probability of a plaintext conditioned on the ciphertext, is equal to the a priori probability of the plaintext, $P(X = x \mid Y = y) = P(X = x)$. Using the mutual information I, this can be expressed as $I(X : Y) = 0$. The mutual information is used to measure both the linear and nonlinear correlation. This is usually difficult to measure but it is a natural measure of the dependence between random

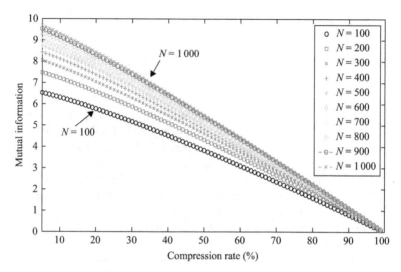

Figure 8.3　Mutual information for an increasing sample length and compression rate

variables, considering the whole dependence structure of these variables [65]. Mutual information is computed as follows:

$$I(X:Y) = \sum_{x \in X} \sum_{y \in Y} p_{xy}(x,y) log_2 \frac{p_{xy}(x,y)}{p_x(x)p_y(y)}, \tag{8.3}$$

where p_{xy} and p_x denote the joint probability density function (PDF) and marginal PDF, respectively.

As (8.1) shows, ciphertext y is a linear projection of plaintext x so, it is expected that no perfect secrecy can be achieved using CS for encryption. For demonstration purposes, we empirically compute the mutual information of a plaintext and its corresponding ciphertext when applying CS, for different plaintext lengths, and for various compression rates. The compression rate is defined as $\frac{N-M}{M} \times 100$, where N and M are the lengths of the plaintext, and the ciphertext, respectively. Observe in Figure 8.3 that as the compression rate increases, mutual information decreases; hence, higher information secrecy is achieved. This is because when the compression rate increases, ciphertext's length becomes smaller, and linear projections are fewer, so the information leakage from the plaintext to the ciphertext reduces. Furthermore, observe that as the length of the plaintext increases, mutual information increases, because for the same compression rate, more information leakage takes place due to the linear projections.

One would assume that by compressing a plaintext using a higher compression rate, would be the ideal solution in a CS crypto-system. Nevertheless, CS performance, in terms of the reconstruction error, deteriorates as compression increases. The performance is also affected by the sparsity of the plaintext. Figure 8.4 shows

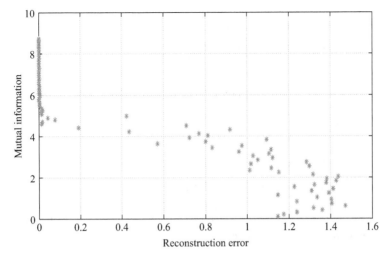

Figure 8.4   Trade-off between the mutual information and the reconstruction error

the trade-off between the mutual information and the reconstruction error e of CS, defined as $e = \frac{||x-\hat{x}||_2}{||x||_2}$, where x and $\hat{x}$ are the original and reconstructed plaintexts, respectively. Compressing a plaintext using a higher compressive rate can provide higher information secrecy, however, the reconstruction error increases.

8.6   Digital signatures

In this section, we will give a quick introduction to the key goals of integrity and authenticity protection on the message level using digital signatures. *Integrity* is violated whenever data or a system is modified in an unauthorized way without being detected. The protection of the integrity of a message against random errors can be fulfilled by a simple checksum like cyclic redundancy check.[1] However, in the presence of an attacker a simple checksum is not sufficient, as an attacker would recalculate the checksum for the maliciously modified message. The mechanism of choice is a so-called protected checksum, that prevents the attacker from being able to recompute the checksum to match malicious changes made [66]. The two most common forms to accomplish this are: keyed hashes, also known as Message Authentication Codes (MAC)[2] and digital signatures. Both can be used to achieve integrity protection against random and malicious errors. To additionally gain a verifiability of origin and also induce accountability for the messages, digital signatures are required, as their base is an asymmetric key pair. Accountability on the level of individual devices

[1] See also ISO 3309.
[2] HMAC is a prominent example see RFC 6151.

is important, consider, e.g. that the home owner and the insurance company might want to know who sent the message that instructed the roof window to open during the heavy rain and caused the living room to be flooded. When the actuator on the window retains a log of the last signed commands, the signature key on the open command can be used to identify the sender.

In this section, we start with briefly defining the overall security goals and then go into the details of the challenges and suggested solutions in the light of the IoT. We conclude with a brief summary.

8.6.1   Goals of integrity protection and origin authentication

The notion of data integrity can be defined as follows:

> ***Data Integrity: The property that data has not been changed, destroyed, or lost in an unauthorized or accidental manner.*** [66].

As a first line of defense, it is important to maintain this state. To run systems as complex as the IoT preventive measures are always good. However, attacks or errors can occur on any involved system or communication, e.g. during the transmission and during storage. So, it is always needed to have a second line of defense to identify if the state of Integrity has been violated. Hence, the protection mechanisms can be further distinguished into preventive and detective measures. The preventive view of Integrity protection (found for example in classic definitions from Clark and Wilson [67]) requires mechanisms that prohibit unauthorized modifications upfront. The latter fulfills its duty as it "detects the violation of internal consistency" [68].

In the remainder, we will concentrate on digital signatures which are a detective measure. Note, that for the correct and expected behavior of the IT system the reason for an unwanted modification does not matter. Altered data must become reliably distinguished from unchanged data. Note, that the Integrity notion and thus also the protection mechanisms for Integrity are not concerned with the quality of the data. If information is incorrectly captured into the system data, integrity protection will prohibit undetected modifications to this false data. This limitation of the protection scope of integrity is important to note from an IoT application's perspective. This situation is covered by the notion of *Veracity* [69] introduced by Gollmann:

> ***Veracity is the property that an assertion truthfully reflects the aspect it makes a statement about.*** [69]

For an IoT example take the case that someone secured all the sensor readings of the smart device with an integrity protection mechanism. Now, errors or attacks that tamper the readings are detected. However, what is not covered is if the sensed value was actually correct. The sensor could, due to a physical error, report wrong values. This data represents false information about the physical entity, hence it does not offer Veracity. However, if no tampering has happened after the wrong value was integrity protected, such the data remains not changed, it still has Integrity.

A digital signature scheme (DSS) is based on asymmetric cryptography. Like in asymmetric encryption, two – related – keys are involved: the secret signature creation key and a related public verification key. The important algorithms of DSS implement the two functionalities: Sign and Verify. The "sign" algorithm takes the secret signing key and the message and generates a signature value. The "verify" algorithm takes the signature value, the message, and the public verification key to obtain the result. If the result is positive, this means that the signature is valid. A secure DSS is correct, meaning that for any correctly signed data the signature over the unchanged data must verify as valid. In turn, this yields two things: First, the data has not been altered according to an integrity policy and second, the signature value has been created involving the secret key corresponding to public key used for verification. To hold the DSS needs to offer unforgeability; EUF-CMA [70] is an adequate model for this.

If a trusted link between an entity (a human user, a device or also a service[3]) and the public verification key exists, a valid signature on a message implies that the message originated at the entity. This gives origin authentication and can be used to build entity authentication protocols [71]. The reason why other methods, not based on asymmetric keys, miss authenticity of origin is that in a symmetric key based integrity check the verifier needs to know the secret that was used to generate the integrity check value.

The most widely known algorithms are ECDSA (based on elliptic curves) and DSA (based on RSA). Hence, in theory the verifier, as well as the "signer", can generate a valid keyed hash for any message.

8.6.2   *Technical challenges and solutions for integrity and origin authentication in the IoT*

Without claiming completeness, we would like to highlight the following challenges, not ordered in any specific way, and give a hint on how they could be solved technically:

- Challenge 1: constrained devices.
- Challenge 2: loosely coupled system (data gets stored in databases, message queues, and the next step is done later).
- Challenge 3: heterogenous interconnected devices need to work smoothly together.
- Challenge 4: setting up keys and trust relationships.
- Challenge 5: privacy.

Note, there might be still more problems, and also the technical solutions proposed here are just meant as a starting point and no complete survey of all existing and appropriate techniques.

---

[3]"System entity – an active part of a system – a person, a set of persons (e.g. some kind of organization), an automated process, or a set of processes – that has a specific set of capabilities." [66].

8.6.2.1   Solution 1: constrained devices require efficient implementations of efficient cryptographic primitives like ECC

Overall, ECC currently presents itself to be usable in software implementations, and is well supported in today's recent IoT hardware platforms [29–31]. ECC was presented already in 1986 independently by Miller [19] and Koblitz [20]. NIST has standard-ized the ECDSA in 2000 [72]. Implementation for WSNs exist, e.g. TinyECC [21], NanoECC [22] or NIST's ECClight [23] and have been further optimized [24]. As another example of how ECC is the perfect candidate for constrained devices, besides Ayuso *et al.*'s work [24], take the 160-bit curve implementation by Kern and Feldhofer on an RFID-type like devices [25] or the very lightweight ECC based construction for authentication of Mayer and Hess [73]. NIST's curves have been criticized [27] and the proposal by Bernstein, which is additionally claimed to be less problematic for implementers, named Curve25519 with the signature algorithm named Ed25519 is currently en route to standardization [28].

In this light, we assume ECC signature generation and verification to be a good choice.

8.6.2.2   Solution 2: loosely coupled system requires message-level integrity

IoT sensory information gathered by the devices and the data is forwarded to other devices or to the backbone servers. In a loosely coupled system, the data is not immediately processed, but often it is stored in message queues to be picked up later by applications to achieve the desired functionality. For example, assume a device with the thermistor to continuously push his readings into a message queue on a cloud server. Asynchronously, this message queue is read by several different applications.

The protection of the integrity for those type of loosely connected systems can be achieved by message-level protection mechanisms. Using cryptographically secure DSSs allows verifying that data was not modified in unauthorized ways. Additionally, you can use a DSS to gain origin-authentication knowing which entity, hence which device, signed the data. Of course, if for privacy reasons this fine-grained identifi-ability is not wanted or permitted, than other cryptographic primitives are needed, e.g. like group-signatures [74].

8.6.2.3   Solution 3: heterogenous interconnected devices need to work smoothly together requires standardizing formats and algorithms

The goal must be to apply integrity protection to the data at the earliest point, i.e. at the device [6], in an end-to-end fashion, all the way downstream to the latest point possibly in the IoT data processing chain. That means that some intermediary devices/nodes might not understand the integrity protection as they see no need to check it. As Figure 8.2 depicts, the MiddleWare or the GateWay might, but need not be the ones who verify the integrity. Hence, this requires that adding security by adding the digital signature to the data, shall not break existing workflows. As from Challenge 2, we already know that the protection of integrity must be on the message level, e.g. each temperature reading carries a signature. So nonsignature aware processing steps for the data in the IoT value chain shall still be able to access the data that contains the signature in the same manner as if it would not be signed.

Assuming that the currently wide spread use of JavaScript Object Notation (JSON) as a transport format, e.g. for CoAP [75], one solution is to add the digital signature as (one or more) additional element(s) into the JSON object that the sensor emits [76]. How the signature gets encoded must be standardized. We would like to point to existing work done in the non-constrained world, e.g. JSON Web Signature (JWS) [77] that is currently discussed in IETF's JOSE working group as an draft. Also you might want to look at CBOR aimed for small code size and small message size described in RFC 7049 [78]. This is used by the IETF's working group COSE [79] to slim down JOSE. To the best of the authors knowledge there is still no agreed solution and future work to do regarding initial ideas,[4] and then the still open standardization of a suitable format for the IoT domain. Also for each curve algorithm[5] the keys must be represented in an interchangeable format. ECC can be put in X.509 certificates, see e.g. [81].

8.6.2.4   Solution 4: setting up keys and trust relationships

As mentioned before, DSSs require asymmetric keys; in other words a pair of public verification key and secret signing key for each device. This requires to generate and manage even more keys than we do today for users and servers. This is a problem that shall not be underestimated, but potentially that is a technical distribution problem that can be managed, e.g. encoding keys into optically scannable QR-codes or transmittable via RFID.

More problematic than the distribution of public keys is to solve the problem of gaining trust into such a public key. A party that verifies a signature using a certain public key shall be convinced that they can deduct trust in a public key. Whether this is done by a hierarchically organized Certificate Issuer, e.g. the CAs in the Public-Key Infrastructure of the Internet today, or more decentralized by a web-of-trust, e.g. like with Pretty Good Privacy, is not important for a secure operation of digital signatures in the IoT. But it must be solved, as without a way to establish trust into the public key used for integrity verification the protection falls to man-in-the-middle attacks. An attacker shall not be able to convince IoT devices that has public key – for which the attacker has the secret signature generation key and thus can produce valid signatures – is trusted to sign firmware updates, or sensitive commands.

As such, the IoT must first bring the trusted keys of central IoT system components, like a firmware update service, onto each and every IoT device. Then they need to be stored securely such that they cannot be exchanged by an attack. The latter requires a secure tamperproof storage solution for the device.

---

[4]An initial not yet length-optimized format was presented in Reference 76.
[5]A list of algorithm descriptors is already defined in Reference 80.

8.6.2.5   Solution 5: privacy for the signing entity by group signature; privacy for data by malleable signatures

The problem of privacy in the IoT does not stop to be important when considering the implementation and integration of integrity protection. This is twofold: On the one hand, the device that generated the signature, by means of a different asymmetric signature generation key unique for each device, may be needed to stay pseudonymous while still being able to check that a certain trusted group of devices had signed it. This can be achieved by a choice of the right cryptographic primitives, e.g. using group signatures. On the other hand, the data that is signed may be subject to changes for the name of privacy but this shall not completely destroy the integrity of the message. This can be achieved using malleable signatures [82]. They allow the removal of specified parts by a specified mandated party (like a delegation) of a signed message without removing the integrity protection. There are manyfold cryptographic schemes, for a comparison of two major strands of those see Reference 83.

8.6.3   *Summary: end-to-end integrity and authenticity based on ECC*

Achieving Integrity for the IoT has two core steps toward its use: (1) adequately lightweight cryptographic primitives to run on the constrained devices, (2) finding and then standardizing efficient signature transport formats and algorithms. ECC currently is the promising and already well supported (in hardware accelerators and software library). Regarding step 2 many groups are discussing this currently, so there will likely be standards to adhere to in the near future. Furthermore, it has adjacent steps that need to be climbed as well: (3) putting trust into public keys and (4) their initial and ongoing distribution. The question of secured storage of keys is the IoT related part and a question regarding the IoT's hardware capabilities, however to induce trust it needs organizational methods, like the secure operation of a CA. Those organizational methods of course can be legally enforced as they can only be technically supported.

The RERUM project has created initial prototypes of an integration of ECC based digital signatures into the IoT ecosystem. Initial results obtained based on ECC digital signature algorithms like NIST's secp256r curve and ed25519 look promising.[6]

8.7   Datagram Transport Layer Security

Securing communication is a far from a trivial task. Difficulties may arise from a wide range of technical or so-called human-factor aspects, including (but not limited to) the design and selection of applicable cryptographic primitives, and the use carefully investigated cryptographic protocols and applications. Moreover, the possible

---

[6]Currently the measurements within RERUM are still running but see Reference 76 for some rough runtime estimations.

designs used to secure communications should not only be investigated in theory, their practical implementations should also be considered. In such a case, a designer should additionally take into consideration, e.g. the efficiency of an algorithm on a vast rage of different platforms and (especially for embedded devices) their protection against implementation attacks [84].

The level of complexity associated with achieving secure communication is already high for an ordinary workstation or a server, and becomes even more challenging in IoT systems. This is mostly due to the fact that new design criteria, such as the power consumption of a device or its computational power, have to be taken into consideration. In many cases, these additional limitations are simply too restrictive to run cryptographic primitives and protocols efficiently.

8.7.1 DTLS protocol for IoT

Probably, one of the best-known and widely deployed cryptographic protocol, which shapes secure communication over the Internet is the Transport Layer Security (TLS) protocol [85]. In general, it allows two parties (called 'peers' or a client and a server) to establish and carry out a secure communication. The TLS protocol is widely used by applications such as Internet browsers, as well as more stream-oriented applications such as voice-over-IP services.

TLS belongs to a group of TCP/IP protocols and (considering the standard OSI model) operates in the session and presentation layers. In fact, within TLS, two protocols named the TLS Record Protocol and TLS Handshake Protocol can be distinguished. The record protocol resides in the fore-mentioned presentation layer and provides secure and reliable connection for the application layer. The handshake protocol, however, resides in the session layer and is mostly responsible for establishing the connection, e.g. for the parameter negotiation and for the authentication of the peers. Since TLS runs over a (reliable) TCP connection, no extra measure on ordering and loosing a packet inside TLS itself is needed.

In general, whenever a high-level protocol uses TCP as an underlying transport protocol, the correct packet order is maintained and no single packet is supposed to be lost during the transmission. This TCP feature is especially important for many cryptographic protocols, where a single packet loss or a wrong packet order might have an impact on the overall message correctness. Unfortunately, the TCP reliability property increases the overheads of the protocol stack, thus in some scenarios more lightweight protocols such as the User Datagram Protocol (UDP) could be used instead. Since UDP features smaller packet headers, requires less memory and has more compact code footprint, it is better suitable for constrained devices which are ubiquitous in IoT systems.

Regrettably, the TLS protocol itself is not applicable (without modification) to UDP connections. To address this issue, a new protocol called DTLS [86] was designed. The new protocol mirrors most of the features and design choices of TLS and extends them with a mechanism to allow packet retransmissions and reordering. This allows the DTLS protocol to run over UDP, providing reliable connection to high-level protocols. Similarly to TLS, DTLS provides authenticity, integrity and confidentiality to the application layer protocols, e.g. it might be used as the underlying protocol to the CoAP [75]. The CoAP is designed especially for resource-constrained environments.

Due to its features, the DTLS protocol for resource-constrained devices has recently gained attention and it is currently investigated by the standardization group 'DTLS In Constrained Environments' (DICE), which is one of the community groups within the IETF. The DICE group aims to specify a DTLS protocol profile suitable for IoT systems [87].

Reassembling TLS, DTLS consists of the handshake and record protocols. The former is used to negotiate a cipher suite, exchange keys, and authenticate the peers, whereas the latter is used to protect traffic data. The handshake protocol uses series of messages (called flights), which are exchanged between communication parties in order to achieve above-mentioned goals. To accomplish these goals, the protocol can select and use public- and private-key schemes. The selection of a suitable mode and credential types has a big impact on the overall handshake performance, thus it is crucial to find an appropriate profile for IoT systems. There are three possible credential types defined by DTLS: a private-key credential type called pre-shared secret, and two public-key credential types called raw public keys and X.509 certificates.

In a pre-shared secret scenario, secret keys and identifiers, which are needed to establish a secure communication between two peers, are deployed and stored in devices (using a trusted channel) before communication. Considering resource-constrained devices, which are likely to be used in IoT, this approach could be, in general, be more efficient – in terms of computation power needed by a device – than a public-key approach. One may also find this solution well-suited to strictly controlled deployment scenarios, where pre-defined connections might make general key management over-engineered. However, in the large scale IoT deployments a pre-shared key solution might not scale enough, and additionally, the large number of keys needing storing might not be suitable for the constrained memory available in such devices.

In contrast to a pre-shared key, in a raw public key scenario, a client and a server use a public/private key pair, which might be generated and further installed on a device by a device manufacturer. To a certain extent this could be seen as a more lightweight version of the X.509 certificates approach, trading off the flexibility provided by X.509 certificates for a better performance. In general, a raw public key credential type might use out-of-band distribution and only a small subset of the X.509 certificate structure. This positively impacts performance and storage requirements, which is essential in IoT systems, but on the other hand limits flexibility provided by X.509 certificates.

The X.509 certificate credential type is the other option (on top of the above-mentioned raw key scenario), where peers use public/private key pair to established communication. To use this mode in the DTLS protocol, a public-key infrastructure, including a Certificate Authority (CA), needs to be deployed. Moreover, devices are augmented with a list of so-called trust anchors to validate certificates. The certificates are deployed in the form specified by the X.509 standard which includes all meta-data information needed. This implies a requirement for increased on-device storage as well as the extra time need to parse a certificate and validate with the CA. Although the X.509 type provides a significant level of flexibility with respect to a key management, its full deployment is usually very challenging considering the limitation of resource-constrained environments.

8.7.2   Summary

Although DTLS seems as a good candidate protocol for standardization aiming to secure communication in IoT systems, its implementation and deployment remains a challenge, mostly due to a constrained nature of resources used in such systems. On the one hand, DTLS provides the possibility to use private-key mode, which could be well-suited in terms of, i.e. computation power, on the other hand this solution suffers from the well-known problem of key management scalability. The solution for a key management issue might be mitigated by using public-key schemes and supporting infrastructure, but public-key operations are costly in general and are still far from being efficient enough for very constrained devices. Advances in EC-based cryptography, including the new EC-based designs such as Curve25519/Ed25519 [28] or the new set of elliptic curves relevant for cryptographic use [88], might bring more suitable solutions for IoT systems.

8.8   Discussion

Security has become a main research area in the IoT world in the last few years. However, the work until now focused only on access control mechanisms for users that need to have access to data that are stored in IoT middlewares, without much concern for the leaf IoT devices. It has been proved though that an IoT system cannot be fully secured without securing the embedded devices themselves. Regardless of how securely the information is sent from the middleware to the applications, the numerous vulnerabilities of the constrained IoT devices leave a whole lot of opportunities to attackers to intercept the data, to hack the devices, to steal personal user information or to affect system decisions. This section presented a high-level framework for increasing the security in the communication layer of IoT systems, with a focus on enhanced lightweight encryption, integrity protection, and overall communication security. The benefits of the proposed framework is that it comprises lightweight mechanisms that can be easily applied on constrained IoT devices and can work either as standalone components or as an integrated security system for increased protection against attacks. Various design choices on where and how to implement these mechanisms were also proposed and the final decision is left to the system administrator, considering the applications that his system supports. This framework can be part of an overall IoT cross-layer security framework, contributing mainly to the protection of the data at the earliest point, something that can significantly increase the overall trustworthiness of the system and IoT in general. Only then, the general public and the business stakeholders can have the necessary incentives for adopting and using or investing in this new set of technologies that can drastically improve the everyday life of people.

References

[1]   M. Weiser, "The computer for the 21st century," *Scientific American*, vol. 265, no. 3, pp. 94–104, 1991.

[2]   Cisco, "The internet of everything. connections counter." Available from: http://newsroom.cisco.com/feature-content?type=webcontent&articleId=1208342, 2013 [Accessed 4 April 2016].

[3]   H. Report, "Internet of things research study, HP report (2014)." Available from: http://www8.hp.com/h20195/V2/GetPDF.aspx/4AA5-4759ENW.pdf, 2014 [Accessed 4 April 2016].

[4]   T. Heer, O. Garcia-Morchon, R. Hummen, S. L. Keoh, S. S. Kumar, and K. Wehrle, "Security challenges in the IP-based internet of things," *Wireless Personal Communications*, vol. 61, no. 3, pp. 527–542, 2011.

[5]   C. M. Medaglia and A. Serbanati, "An overview of privacy and security issues in the internet of things," *The Internet of Things*. Berlin: Springer, 2010, pp. 389–395.

[6]   H. C. Pöhls, V. Angelakis, S. Suppan, *et al.*, "RERUM: Building a reliable IoT upon privacy- and security-enabled smart objects," in *Wireless Communications and Networking Conference Workshops (WCNCW), 2014 IEEE*. Piscataway, NJ: IEEE, 2014, pp. 122–127.

[7]   O. Vermesan and P. Friess, *Building the Hyperconnected Society. Internet of Things Research and Innovation Value Chains, Ecosystems and Markets*. River Publishers: Aalborg, Denmark, 2015.

[8]   A. Bassi, M. Bauer, M. Fiedler, *et al.*, *Enabling Things to Talk: Designing IoT Solutions with the IoT Architectural Reference Model*. Berlin: Springer, 2013.

[9]   S. Ménoret, "iCore Internet Connected Objects for Reconfigurable Ecosystem: D2.5 Final architecture reference Model". iCore project deliverable available from: http://cordis.europa.eu/docs/projects/cnect/8/287708/080/deliverables/001-20141031finalarchitectureAres20143821100.pdf, 2014 [Accessed 20 June 2016].

[10]   Butler Consortium, "D3.2 Integrated System Architecture and Initial Pervasive BUTLER proof of concept". A BUTLER project deliverable available from: http://www.iot-butler.eu/wp-content/uploads/downloads/2013/10/D3.2-Integrated-System-Architecture-v1.50.pdf, 2013 [Accessed 20 June 2016).

[11]   D. Rotondi, S. Piccione, G. Altomare, *et al.*, "IoT@Work: D1.3 – Final framework architecture specification." IoT@Work project deliverable available from: https://www.iot-at-work.eu/data/D1.3_IoT@Work_Architecture_final_v1.0-submitted.pdf, 2013 [Accessed 20 June 2016].

[12]   P. Dimitropoulos, "OpenIoT: D2.3 Architecture and proof-of-concept specifications." OpenIoT project deliverable, 2012.

[13]   F. Carrez, G. Kousiouris, A. Marinakis *et al.*, "COSMOS: D2.3.2 Conceptual model and reference architecture." A COSMOS project deliverable available from: http://iot-cosmos.eu/sites/default/files/cosmos/files/content-files/deliverables/Cosmos_D2.3.2_Conceptual%20Model%20and%20

Reference%20Architecture%20(Updated)_v1.0_0.pdf, 2015 [Accessed 20 June 2016].

[14] XILINX. "Xilinx Zynq-7000 all programmable SoC ZC702 evaluation kit." Available from: http://www.xilinx.com/products/boards-and-kits/ek-z7-zc702-g.html, 2016 [Accessed 20 June 2016].

[15] A. Skarmeta, A. Quesada, R. Marín Pérez *et al.*, "SMARTIE: D2.3 Initial architecture specification." A SMARTIE project deliverable available from: http://cordis.europa. eu/docs/projects/cnect/2/609062/080/deliverables/001-D23Ares2015553314. pdf, 2015 [Accessed 20 June 2016].

[16] A. Orsdemir, H. Altun, G. Sharma, and M. Bocko, "On the security and robustness of encryption via compressed sensing," in *Proceedings of MILCOM*. San Diego, California, USA, 2008, pp. 1–7.

[17] Y. Rachlin and D. Baron, "The secrecy of compressed sensing measurements," in *Proceedings of Allerton Conference on Communication, Control, and Computing*. Monticello, IL, USA, 2008, pp. 813–817.

[18] V. Cambareri, M. Mangia, F. Pareschi, R. Rovatti, and G. Setti, "Low-complexity multiclass encryption by compressed sensing," *IEEE Transactions on Signal Processing*, vol. 63, pp. 2183–2195, 2015.

[19] V. Miller, "Use of elliptic curves in cryptography," in *Advances in Cryptology – CRYPTO '85 Proceedings*. Berlin: Springer, 1986, pp. 417–426.

[20] N. Koblitz, "Elliptic curve cryptosystems," *Mathematics of Computation*, vol. 48, no. 177, pp. 203–209, 1987.

[21] A. Liu and P. Ning, "Tinyecc: A configurable library for elliptic curve cryptography in wireless sensor networks," in *Information Processing in Sensor Networks, 2008. IPSN '08. International Conference on*. Piscataway, NJ: IEEE, 2008, pp. 245–256.

[22] P. Szczechowiak, L. B. Oliveira, M. Scott, M. Collier, and R. Dahab, "NanoECC: Testing the limits of elliptic curve cryptography in sensor networks," *Wireless Sensor Networks*. Berlin: Springer, 2008, pp. 305–320.

[23] National Institute of Standards and Technology (NIST), "ecc-light-certificate library." Available from: https://github.com/nist-emntg/ecc-light-certificate, 2014 [Accessed 4 April 2016].

[24] J. Ayuso, L. Marin, A. Jara, and A. F. G. Skarmeta, "Optimization of public key cryptography (RSA and ECC) for 16-bits devices based on 6lowpan," in *First International Workshop on the Security of the Internet of Things*, Tokyo, Japan, 2010.

[25] T. Kern and M. Feldhofer, "Low-resource ECDSA implementation for passive RFID tags," in *17th IEEE International Conference on Electronics, Circuits, and Systems (ICECS'10)*. Piscataway, NJ: IEEE, 2010, pp. 1236–1239.

[26] M. Braun, E. Hess, and B. Meyer, "Using elliptic curves on RFID tags," *International Journal of Computer Science and Network Security*, vol. 2, pp. 1–9, 2008.

[27] D. J. Bernstein, T. Chou, C. Chuengsatiansup, *et al.*, "How to manipulate curve standards: A white paper for the black hat," *Cryptology ePrint Archive*, Report 2014/571, Tech. Rep., 2014.

[28]   N. Moeller and S. Josefsson, "IETF draft: EdDSA and Ed25519." Available from: https://tools.ietf.org/html/draft-josefsson-eddsa-ed25519-02, Feb. 2015 [Accessed 4 April 2016].

[29]   Atmel Corporation, "ECC-based devices." Available from: http://www.atmel.com/products/security-ics/cryptoauthentication/ecc-256.aspx, 2015 [Accessed 4 April 2016].

[30]   Openmote.org, "Openmote CC2538." Available from: http://www.openmote.com/hardware/openmote-cc2538-en.html, 2014 [Accessed 4 April 2016].

[31]   Zolertia, "Zolertia ReMote." Available from: http://zolertia.io/product/hardware/re-mote, 2015 [Accessed 4 April 2016].

[32]   E. Z. Tragos, V. Angelakis, A. Fragkiadakis, *et al.*, "Enabling reliable and secure IoT-based smart city applications," in *Pervasive Computing and Communications Workshops (PERCOM Workshops), 2014 IEEE International Conference on*. Piscataway, NJ: IEEE, 2014, pp. 111–116.

[33]   E. J. Candè and M. B. Wakin, "An introduction to compressive sampling," *Signal Processing Magazine, IEEE*, vol. 25, no. 2, pp. 21–30, 2008.

[34]   D. E. Fu and J. A. Solinas, "Ike and ikev2 authentication using the elliptic curve digital signature algorithm (ECDSA)," 2007.

[35]   D. Boneh, X. Boyen, and H. Shacham, "Short group signatures," *Advances in Cryptology – CRYPTO 2004*. Berlin: Springer, 2004, pp. 41–55.

[36]   H. C. Pöhls, S. Peters, K. Samelin, J. Posegga, and H. de Meer, "Malleable signatures for resource constrained platforms," *Information Security Theory and Practice. Security of Mobile and Cyber-Physical Systems*. Berlin: Springer, 2013, pp. 18–33.

[37]   P. Charalampidis, A. G. Fragkiadakis, and E. Z. Tragos, "Rate-adaptive compressive sensing for IoT applications," in *Vehicular Technology Conference (VTC Spring), 2015 IEEE 81st*. Piscataway, NJ: IEEE, 2015, pp. 1–5.

[38]   V. Kumar, G. Oikonomou, T. Tryfonas, D. Page, and I. Phillips, "Digital investigations for IPv6-based wireless sensor networks," *Digital Investigation*, vol. 11, Supplement 2, pp. S66–S75, August 2014.

[39]   G. Montenegro, N. Kushalnagar, J. W. Hui, and D. E. Culler, "Transmission of IPv6 packets over IEEE 802.15.4 networks," RFC 4944, Sep. 2007.

[40]   J. Hui and P. Thubert (Eds.), "Compression format for IPv6 datagrams over IEEE 802.15.4-based networks," RFC 6282, Sep. 2011.

[41]   Z. Shelby, S. Chakrabarti, E. Nordmark, and C. Bormann (Eds.), "Neighbor discovery optimization for low-power and lossy networks," RFC 6775, Nov. 2012.

[42]   T. Winter, P. Thubert, A. Brandt, *et al.* (Eds.), "RPL: IPv6 routing protocol for low power and lossy networks," RFC 6550, Mar. 2012.

[43]   P. Nikander, J. Kempf, and E. Nordmark (Eds.), "IPv6 neighbor discovery (ND) trust models and threats," RFC 3756, May 2004.

[44]   A. Alsa'deh and C. Meinel, "Secure neighbor discovery: Review, challenges, perspectives, and recommendations," *IEEE Security Privacy*, vol. 10, no. 4, pp. 26–34, Jul. 2012.

[45]   A. Le, J. Loo, A. Lasebae, M. Aiash, and Y. Luo, "6LoWPAN: A study on QoS security threats and countermeasures using intrusion detection system approach," *International Journal of Communication Systems*, vol. 25, no. 9, pp. 1189–1212, Sep. 2012.

[46]   A. Dvir, T. Holczer, and L. Buttyan, "VeRA – Version number and rank authentication in RPL," in *2011 IEEE Eighth International Conference on Mobile Ad Hoc and Sensor Systems (MASS)*. Valencia, Spain, October 2011, pp. 709–714.

[47]   T. Tsao, R. K. Alexander, M. Dohler, V. Daza, A. Lozano, and M. Richardson (Eds.), "A security threat analysis for routing protocol for low-power and lossy networks (RPL)," Internet Draft (version 06), December 2013 (draft-ietf-roll-security-threats).

[48]   L. Wallgren, S. Raza, and T. Voigt, "Routing attacks and countermeasures in the RPL-based Internet of Things," *International Journal of Distributed Sensor Networks*, vol. 13, no. 794326, 2013, 11pp.

[49]   S. Raza, L. Wallgren, and T. Voigt, "SVELTE: Real-time intrusion detection in the Internet of Things," *Ad Hoc Networks*, vol. 11, no. 8, pp. 2661–2674, Nov. 2013.

[50]   T. Clausen and U. Herberg, "Some considerations on routing in particular and lossy environments," in *Proceedings of the First Interconnecting Smart Objects with the Internet Workshop*. Prague, Czech Republic, Mar. 2011.

[51]   P. Ilia, G. Oikonomou, and T. Tryfonas, "Cryptographic key exchange in IPv6-based low power, lossy networks," in *Proceedings of the Workshop in Information Theory and Practice (WISTP 2013)*, series in *Lecture Notes in Computer Science*, vol. 7886. Berlin: Springer, May 2013, pp. 34–49.

[52]   T. O'Donovan, J. Brown, F. Büsching, *et al.*, "The GINSENG system for wireless monitoring and control: Design and deployment experiences," *ACM Transactions on Sensor Networks*, vol. 10, no. 1, pp. 4:1–4:40, Dec. 2013.

[53]   E. Byres and J. Lowe, "The myths and facts behind cyber security risks for industrial control systems," in *Proceedings of the VDE Kongress*, vol. 116. Berlin, Germany, 2004.

[54]   B. Triki, S. Rekhis, and N. Boudriga, "Digital investigation of wormhole attacks in wireless sensor networks," in *Proceedings of the Eighth IEEE International Symposium on Network Computing and Applications (NCA 2009)*. Cambridge, MA, USA, 2009, pp. 179–186.

[55]   F. Mouton and H. Venter, "A prototype for achieving digital forensic readiness on wireless sensor networks," in *Proceeding of the AFRICON, 2011*. Livingstone, Zambia, 2011, pp. 1–6.

[56]   S. Rekhis and N. Boudriga, "Pattern-based digital investigation of x-hole attacks in wireless ad hoc and sensor networks," in *Proceedings of the International Conference on Ultra Modern Telecommunications & Workshops (ICUMT'09)*. St Petersburg, Russia, 2009, pp. 1–8.

[57]   S. Dawans and L. Deru, "Demo abstract: Foren6, a RPL/6LoWPAN diagnosis tool," in *Proceedings of the 11th European Conference on Wireless Sensor Networks (EWSN)*. Oxford, UK, Feb. 2014.

[58]   B. Kaliski, "The mathematics of the RSA public-key cryptosystem," *RSA Laboratories*. Available from: http://www.mathaware.org/mam/06/Kaliski.pdf [Accessed 20 June 2016].

[59]   K. Raeburn, "Advanced encryption standard (AES) encryption for kerberos 5," *RFC 3962*, 2005.

[60]   G. Werner, K. Lorincz, M. Ruiz, *et al.*, "Deploying a wireless sensor network on an active volcano," *IEEE Internet Computing, Special Issue on Data-Driven Applications in Sensor Networks*, vol. 10, pp. 18–25, 2006.

[61]   A. Dunkels, "The contikimac radio duty cycling protocol," Swedish Institute of Computer Science, Technical Report, 2011.

[62]   A. Fragkiadakis, I. Askoxylakis, and E. Tragos, "Secure and energy-efficient life-logging in wireless pervasive environments," in *Proceedings of the First International Conference on Human Aspects of Information Security, Privacy and Trust*. Las Vegas, NV, USA, 2013, pp. 306–315.

[63]   E. Candes and M. Wakin, "An introduction to compressive sampling," *IEEE Signal Processing Magazine*, vol. 25, no. 2, pp. 21–30, 2008.

[64]   C. Shannon, "Communication theory of secrecy systems," *Bell System Technical Journal*, vol. 28, pp. 656–715, 1949.

[65]   J. Williams and Y. Li, *Estimation of Mutual Information: A Survey*. Rough Sets and Knowledge Technology, *Lecture Notes in Computer Science*. Berlin: Springer, 2009.

[66]   R. Shirey, "Internet security glossary, version 2." IETF RFC 4949, 2007.

[67]   D. D. Clark and D. R. Wilson, "A comparison of commercial and military computer security policies," *Security and Privacy, IEEE Symposium on*, vol. 0, pp. 184–194, 1987.

[68]   E. Michiels, "ISO/IEC 10181-6: 1996 Information technology – Open Systems Interconnection – Security frameworks for open systems: Integrity framework," *ISO Geneve*, Switzerland, 1996.

[69]   D. Gollmann, "Veracity, plausibility, and reputation," *Information Security Theory and Practice. Security, Privacy and Trust in Computing Systems and Ambient Intelligent Ecosystems*, Berlin, Heidelberg, Germany: Springer-Verlag, 2012, pp. 20–28.

[70]   S. Goldwasser, S. Micali, and R. L. Rivest, "A digital signature scheme secure against adaptive chosen-message attacks," *SIAM Journal on Computing*, vol. 17, pp. 281–308, 1988.

[71]   D. Gollmann, "What do we mean by entity authentication?" in *Security and Privacy, 1996. Proceedings, 1996 IEEE Symposium on*. Piscataway, NJ: IEEE, 1996, pp. 46–54.

[72]   National Institute of Standards and Technology (NIST), "PUB FIPS 186-4. Digital signature standard (DSS)," NIST: Gaithersburg USA, 2011.

[73]   B. Meyer and E. Hess, "United States Patent 8,850,213." Available from: http://www.uspto.gov/web/patents/patog/week39/OG/html/1406-5/ US08850213-20140930.html, Jul. 2014 [Accessed 4 April 2016].

[74]   D. Chaum and E. Van Heyst, "Group signatures," in *Proceedings of the 10th Annual International Conference on Theory and Application of Cryptographic Techniques*. Berlin: Springer-Verlag, 1991, pp. 257–265.

[75]   Z. Shelby, K. Hartke, and C. Bormann, "The constrained application protocol (CoAP)," *RFC 7252*, Jun. 2014.

[76]   H. C. Pöhls, "JSON sensor signatures (JSS): End-to-end integrity protection from constrained device to IoT application," in *Innovative Mobile and Internet Services in Ubiquitous Computing (IMIS), 2015 Ninth International Conference on*. Piscataway, NJ: IEEE, 2015, pp. 306–312.

[77]   M. Jones, J. Bradley, and N. Sakimura, "IETF draft: JSON Web Signatures (JWS)." Available from: https://tools.ietf.org/html/draft-ietf-jose-json-web-signature-41, Jan. 2015 [Accessed 4 April 2016].

[78]   C. Bormann and P. Hoffman, "Concise Binary Object Representation (CBOR)," *RFC 7049*, Internet Engineering Task Force, 2013 [Online]. Available from: https://tools.ietf.org/html/rfc7049 [Accessed 4 April 2016].

[79]   C. Bormann, "IETF draft: CBOR Object Signing and Encryption (COSE)." Available from: https://tools.ietf.org/html/draft-bormann-jose-cose-00, Oct. 2014 [Accessed 4 April 2016].

[80]   M. Jones, "IETF draft: JSON Web Algorithms (JWA)." Available from: https://tools.ietf.org/html/draft-ietf-jose-json-web-algorithms-40, Jan. 2015 [Accessed 4 April 2016].

[81]   S. Blake-Wilson, D. Brown, and P. Lambert, "Use of elliptic curve cryptography (ECC) algorithms in cryptographic message syntax (CMS)," Tech. Rep., 2002.

[82]   H. C. Pöhls and M. Karwe, "Redactable signatures to control the maximum noise for differential privacy in the smart grid," in *Proceedings of the Second Workshop on Smart Grid Security (SmartGridSec 2014)*, series in *Lecture Notes in Computer Science (LNCS)*, J. Cuellar (Ed.), vol. 8448. Berlin: Springer International Publishing, 2014.

[83]   H. de Meer, H. C. Pöhls, J. Posegga, and K. Samelin, "On the relation between redactable and sanitizable signature schemes," in *ESSoS*, series in *LNCS*, vol. 8364. Berlin: Springer, 2014, pp. 113–130.

[84]   P. Kocher, J. Jaffe, and B. Jun, "Differential power analysis," in *Advances in Cryptology – CRYPTO'99*. Berlin: Springer, 1999, pp. 388–397.

[85]   T. Dierks and E. Rescorla, "The transport layer security (TLS) protocol version 1.2," *RFC 5246*, Aug. 2008.

[86]   N. Modadugu and E. Rescorla, "The design and implementation of datagram TLS," in *NDSS*. San Diego, California, USA, 2004.

[87]   Internet Engineering Task Force (IETF), "TLS/DTLS profiles for the internet of things." Available from: https://tools.ietf.org/pdf/draft-ietf-dice-profile-17.pdf, Oct. 2015 [Accessed 4 April 2016].

[88]   J. W. Bosand, C. Costello, P. Longa, and M. Naehrig, "Selecting elliptic curves for cryptography: An efficiency and security analysis," *Journal of Cryptographic Engineering*, pp. 1–28, 2014.

Chapter 9

Lightweight cryptographic identity solutions for the Internet of Things

Chongyan Gu, Neil Hanley, and Máire O'Neill

Summary

With the increasing emergence of pervasive electronic devices in our lives, the Internet of Things (IoT) has become a reality with its influence on our day to day activities set to further increase with a projected 50 billion connected devices by the year 2020 [8]. These smart devices and sensors will be found in our homes, our cars, our workplaces, etc., and have the potential to revolutionise how we interact with the world today. The slew of data generated by such a volume of devices necessitates the use of smart, autonomous machine-to-machine (M2M) communications; however, this necessarily poses serious security and privacy issues as we will no longer have direct control over with whom and what our devices communicate. This could potentially open up new attack vectors for criminal hackers to exploit through the use of malicious or tampered IoT devices. Compounding the problem is that to enable the ubiquitous nature of the IoT, the embedded devices themselves are often low-cost, low-power, throwaway units which are restricted both in memory and computing power. Generally, low-cost devices targeted at the IoT space, such as the ARM Cortex-M® or the Atmel tinyAVR® families of microcontroller units (MCUs), contain little if any embedded security features. Their lightweight nature is such that even highly optimised cryptographic implementations targeted at specific MCU still require a significant timing, and corresponding energy, overhead [9]. Hence, it is clear we need a new approach to securing the IoT. In this chapter, we outline the proposed use of Physical Unclonable Functions (PUFs) for the provision of IoT device security.

9.1 Introduction

Physical unclonable functions (PUFs) are a novel security primitive which utilise the variation that occurs during manufacturing processes in order to generate a unique intrinsic identifier or challenge response set for a device. First proposed by Naccache and Fremanteau *et al.* [33] as part of an Identification (ID) scheme for smart-cards, it was not until the publications by Pappu *et al.* [35], and Gassend *et al.* [16] that the academic and security communities began to investigate the concept further. Earlier proposals often required bulky external measurement equipment in order to extract the PUF reading, however the proposal by Gassend *et al.* [16] introduced the concept of an integrated *silicon* PUF where the measurement circuitry is contained within the silicon itself allowing for direct integration with application specific integrated circuit (ASIC), enabling straightforward incorporation for security purposes. Subsequent research also saw PUF constructions tailored for field programmable gate arrays (FPGAs) [25] and MCUs [30], encompassing the main underlying computational platforms utilised in the IoT space. A PUF can be viewed as a circuit that uses manufacturing process variations to generate a unique digital fingerprint for a device, making it resistant to cloning even by an untrusted fabrication facility with full knowledge of the design specifications. Such a primitive has a number of desirable properties from a security aspect, such as the ability to provide low-cost unique identification for an Integrated Circuit (IC) or to provide a *variability aware* circuit that returns a device specific response to an input challenge. This gives some advantages over current state-of-the-art alternatives such as secure or non-volatile memory (NVM), or trusted platform modules' (TPM's). No special manufacturing processes are required to integrate a PUF into a design lowering the overall cost of the manufacturing process, and everything can be kept on-chip enabling the PUF to be utilised as a hardware root of trust for all security or identity related operations on the device. This enables a multitude of higher level operations based on secure key storage and chip authentication, as well as potential anti-counterfeiting capabilities for supply chain security.

This chapter is split into two main topics: (1) Related work and PUF design considerations and (2) an evaluation of an Identity (ID) based PUF for IoT applications. Initially, we outline the underlying concept of a PUF, and give an overview of some of the various approaches to constructing one, as well as a description of some suitable metrics to enable a fair comparison between PUF designs. This is followed by an outline of various error correction and control mechanisms, as well as when they are required to enable higher level protocols. The chapter will discuss other design considerations, such as modelling and side-channel attacks (SCAs), which also need to be taken into account when designing secure PUF based systems.

The second topic of this chapter will consist of a PUF evaluation for an ID based IoT scenario. A target device for this kind of application could be the Xilinx® Zynq-7000 [47] System-on-Chip (SoC) platform series, which contains a dual-core ARM® Cortex-A9 MCU, combined with either a low-cost Artix or medium-power Kintex FPGA fabric on a single chip. This gives the speed and flexibility of software programming to handle the communication and protocol stacks, while having the computational power of a tightly coupled FPGA enabling real-time analysis of large volumes of data. This might be viewed as an atypical IoT device, however it is rapidly gaining traction in embedded scenarios, and can be found in applications where performance is of greater importance than power consumption or cost, such as in real-time decision making for autonomous transportation or smart networks.

9.2 Related work

PUFs are a promising new security primitive for electronic devices, with a number of new companies such as Verayo [3], Intrinsic-ID [1], and Sirona [2], now beginning to provide commercial solutions based on the technology, as the promise of a unique, unclonable identity for an electronic device is attractive from a security viewpoint.

All electronic devices contain some inherent variability that is unique to a particular instance of a particular device. This is due to the underlying random processes that occur during fabrication. A PUF looks to utilise this randomness in order to create a unique *digital fingerprint* for the device. This fingerprint must contain sufficient entropy such that an adversary, given an arbitrary number of similarly programmed devices, cannot learn enough information which would allow him to attack, as yet, an unseen device. The unclonability aspect comes from the fact that as the random process variations on which the PUF is based cannot be controlled, a physical clone is impossible.

The general usage of a PUF is that for a given input challenge, the PUF circuitry will return a unique response for each device as can be seen in Figure 9.1. One design approach (delay-based approach) to utilise this randomness is through the creation of configurable (via the input challenge) *race* conditions between two signals with the output bit determined by which signal arrives first. This includes proposals such as the Arbiter PUF [16] and its variants, such as the XOR – Arbiter PUF [41]. A second approach involves memory based designs such as the static random-access memory (SRAM) [19] PUF based on the start-up values of random access memory (RAM), or the Butterfly [25] PUF which is based on cross-coupled NAND gates. The third main design approach is through the use of ring oscillators (ROs) and checks which of a randomly selected RO pair has reached the higher frequency value in a fixed time period [16]. Each approach has its own time/area/entropy trade-offs, and it is likely that, ultimately, designs will be chosen to fit a particular application.

It must also be noted that entropy levels for various designs vary greatly depending on their underlying complexity. In order to distinguish between the security levels of intrinsic PUF designs, Guajardo *et al.* [19] introduced two sub-categories with regard

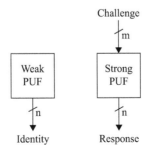

Figure 9.1 Weak and strong PUF concepts

to the use of challenge response pairs (CRPs), namely, *weak* PUFs and *strong* PUFs. The denomination *weak* here does not imply that the design is weak from a security perspective; it refers to the fact that they can only be used for ID generation and not for CRP-based protocols where the outputs might be externally visible, as they generally do not contain sufficient entropy to prevent an adversary building a software model of the PUF given access to enough responses.

9.2.1   Weak PUFs

Weak PUFs exhibit the following characteristics; firstly they may have very few or no challenges. In the extreme case, a *weak* PUF generates just a single response. It is assumed that an attacker cannot access the response of *weak* PUFs as one or a few CRPs would be sufficient to build a model of the security system.

As mentioned previously, two example *weak* PUF designs are the SRAM PUF, proposed by Guajardo *et al.* [19] and the Butterfly PUF presented by Kumar *et al.* [25]. Typical uses for this class of PUF are intrinsic key generation in place of secure memory, or seed generation for a pseudorandom number generator (PRNG). No special manufacturing process is required to implement the PUF, and when compared to other key storage approaches involving NVM, it is difficult to reverse-engineer or derive the output of a *weak* PUF since the keys are only generated when required and the device is powered-up, hence offline tampering attacks are largely infeasible to all but the most determined adversary [42].

9.2.2   Strong PUFs

In contrast to *weak* PUFs, *strong* PUFs have very many possible CRPs which can be used for lightweight mutual authentication protocols. To be classed as a *strong* PUF, the design must meet the following properties [19]:

- The mapping between challenge and response is unique and independent for each CRP, and between different PUF instances, as well as being unknown to the device manufacturer.
- The CRP set size must be large enough that an attacker cannot simply enumerate all possible options to create a look up table that would emulate the PUF.
- The response must contain sufficient entropy such that it is computationally hard to create a model that can (partly) determine the relationship between the challenges and responses.

Strong PUFs share almost all the advantages of *weak* PUFs. However, the architecture of *strong* PUFs has a more complex challenge-response behaviour than *weak* PUFs and hence has independent responses for different challenges, the trade-off being greater hardware requirements and/or response time. The Arbiter PUF proposed by Gassend *et al.* [16] is a typical *strong* PUF. A review of *strong* PUFs and corresponding authentication proposals is provided by Delvaux *et al.* [11, 12].

While there have been numerous *strong* PUF proposals in the academic literature, in practice the requirements as listed above are difficult to meet for a reasonable hardware overhead as subtle relationships often exist between the CRPs. Given an error free *weak* PUF, it can be used in the guise of a *strong* PUF in conjunction with a cryptographically secure block cipher or some keyed one-way function (OWF), with the properties of the cryptographic function ensuring that there is no relationship between input and output. Conversely, a *strong* PUF can also be used as a *weak* PUF by fixing the input challenge, however given that generally *strong* PUFs are noisier and have greater hardware requirements, the benefits of this are fewer.

9.2.3 Evaluation metrics

In order to evaluate the suitability of a PUF design for integration into a security product, extensive testing must be conducted to verify its operation. As it is random manufacturing process variations that PUFs seek to utilise, simulation, while valuable, cannot fully capture the true nature of the PUF design and it needs to be implemented on the target hardware device for evaluation (e.g. see the 250 FPGA test-bed setup built as part of the FP7 SPARKS project [17]). The two main metrics used to evaluate designs are uniqueness and reliability as outlined below. These are not the only metrics however, and prior to deployment the PUF should also be tested for robustness, uniformity, bit-aliasing and entropy [4, 28].

9.2.3.1 Uniqueness

The uniqueness of a PUF determines how easily a particular PUF design can be used to differentiate between different devices. Specifically, it quantifies the average inter-chip Hamming distance (HD), HD_{inter}, between sets of responses extracted from different devices, which implement the *same* PUF design, and have been supplied with the *same* challenge in order to show the extent by which the responses of the chips differ. Accordingly, a normalized measure for uniqueness based on HD_{inter} can be defined as follows. Given two PUF instances P_i and P_j that implement the same ID generator and have *n*-bit responses R_i and R_j given some challenge, C, then the uniqueness, expressed as HD_{inter} among k devices, is defined as in (9.1), which is the normalized pairwise HD between each pair of responses. For a suitable random PUF design, the HD_{inter} value should be approximately 50%.

$$\text{Uniqueness} = HD_{inter} = \frac{2}{k(k-1)} \sum_{i=1}^{k-1} \sum_{j=i+1}^{k} \frac{HD(R_i, R_j)}{n} \times 100 \qquad (9.1)$$

9.2.3.2 Reliability

For many key storage and key generation applications, it is required that the output of a PUF design can be perfectly reproduced. However, fluctuations in supply voltage and ambient temperature cause noise on the response bits. Therefore, reliability is used as a percentage measure of the number of noisy response bits to quantify the error in the response, when the PUF design is subjected to environmental variations. Designs yielding a higher error rate will require additional error correction circuitry. For a device *i*, reliability is established as a single value by finding the intra-chip HD, HD_{intra}, of *m* *n*-bit response samples, R_i', taken at different supply voltages and temperatures compared to a baseline reference response, R_i, taken at nominal operating conditions. Therefore, for a given device, HD_{intra} is defined as in (9.2).

$$HD_{intra} = \frac{1}{s} \sum_{t=1}^{s} \frac{HD(R_i, R'_{i,t})}{n} \times 100 \qquad (9.2)$$

The percentage measure for reliability can be found in (9.3), and can be applied in situations of varying supply and ambient temperature to test how the PUF operates in conditions it might experience in the field. Ideally this reliability value should be close to 100%.

$$\text{Reliability} = 100 - HD_{intra} \qquad (9.3)$$

9.2.4 Attack scenarios

While PUFs are promising cryptographic primitives for securing lightweight devices, they are not immune from some traditional attacks if designed poorly, or not integrated correctly into a higher level protocol. For instance, in a naïvely implemented CRP protocol where the challenges are responses are sent in the clear, if they are not discarded after a single use this could open up a trivial replay attack to an adversary monitoring the communications. While this particular attack scenario is easily avoided, there are many other possible subtle attacks based on modelling PUF circuits.

Machine learning based modelling attacks can be particularly effective against many PUF designs.[1] For example, many arbiter designs can be modelled as additive delay circuits and linear regression used to estimate the resultant delay units. This weakness was known from the early days of PUF design proposals; Support Vector Machines (SVMs) were used to model a PUF in Reference 16, and a more recent overview can be found in References 37, 38. A practical modelling attack against commercial PUF tags has also been demonstrated in Reference 6. Another closely related attack that can be effective is an application of traditional differential and linear cryptanalysis [15]. In this research the authors, rather than choosing random challenges to obtain responses to build a model, sent low HD related challenges in order to determine what, if any, parts of the PUF circuitry had a disproportionate effect on the output as a whole. Hence, it is clear that PUFs, while still promising, must be carefully designed when implementing for commercial applications.

Physical implementation attacks such as power and electromagnetic based SCAs, have also been conducted against PUF implementations with varying degrees of success [7, 31, 32]. These attacks are of particular concern for embedded systems and there have been many examples of attacks on real-world devices [24, 34]. When attacking a PUF based system in this way, often the post-processing or error correction circuitry (ECC) can be more susceptible than the PUF itself; however it was shown in Reference 42 that delay times in an arbiter PUF can be estimated to an accurate enough degree using photonic emission analysis.

---

[1]Note that this is more of an issue for *strong* PUF designs as *weak* PUF outputs should never be externally visible.

9.2.5 Protocol level options

The term *Internet of Things* encompasses a wide variety of devices and projected applications. For example, a connected home scenario might have a number of low-cost heat and humidity sensors communicating through a central gateway to a home heating unit, which in turn is communicating with a smart meter connected to the utility base station (via a mesh network). The base station may be receiving on-demand pricing based on the current draw on the regional electricity network, in order to determine the optimal time to heat the home prior to when the inhabitants return after work, which in turn is determined in real time over the cellular network via an application on a mobile phone or connected car. This arbitrary example shows the potential the IoT has to simplify and improve our lives, however it also shows the increased attack entry points for malicious adversaries to obtain detailed knowledge of our lives. Presently the takeover of any of the multitude of IoT devices requires strong cryptographic and privacy preserving protocols, which can be problematic on the low-cost devices projected to make up the bulk of the IoT.

Dependent on the application, it is anticipated that PUF can be used to replace or augment computationally intense public-key infrastructure (PKI) systems. Assuming the scenario of a constrained client device, and a resource rich server as might be found in a typical system, the naïve PUF authentication protocol as previously mentioned could be used to allow the server authenticate the client [35]. Assuming a *strong* PUF with sufficient entropy on the client side (and it has not yet been shown that such a PUF exists), during enrolment prior to deployment, the server, using a true random number generator (TRNG), can simply read back enough CRPs in order to authenticate the client for the duration of its lifetime. During authentication, the server then queries the client with a stored challenge, and if the response is within a certain HD of the stored response then the client is accepted as legitimate. The accepted error can be determined during enrolment by an analysis of the acquired CRP dataset. However as each CRP can only be used once, dependent on the target application this can lead to significant storage requirements, particular where there is a large number of client devices, with enrolment time also likely to be an issue.

As the existence of a *strong* PUF is not yet fully proven, their use for authentication purposes requires them to be used in a controlled manner as shown in Figure 9.2 [16]. Due to the use of hash functions, etc., single-bit errors quickly propagate hence can no longer be tolerated, and correction mechanisms such as fuzzy extractors must be employed [14], which requires additional information to be stored for every CRP. However this detracts from the *lightweight* advantage of using a PUF in the first place, and as this helper data is public, it is possible that secret or privacy revealing information could be leaked as messages are being exchanged. Protocols such as *reverse fuzzy extractors* [22] have been suggested that move the computational workload of error correction to the server side, however an analysis of *strong* PUF protocols [10, 12] found that none of the proposed protocols fulfils all of the required security and resource requirements.

Currently, *weak* PUFs used for secret key generation are an attractive proposition for mutual authentication protocols. In this scenario the server side only needs to store a single value per client, and mutual authentication can be achieved through the

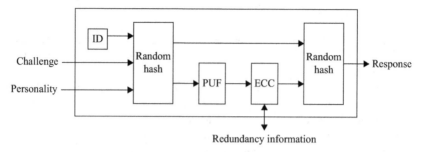

Figure 9.2   Controlled PUF

encryption of a random value with a lightweight cipher using the PUF identity output as a key. The proviso here is that the output must be error free on every instantiation. Derivatives of this key can also be used for other cryptographic functions such as PRNG seeding or to bind a PKI certificate to a particular piece of hardware. An example of an end-to-end design of a PUF-based authentication protocol can be found in Reference 5.

9.2.6   Error correction

If the underlying PUF has a low error rate, simple majority voting or selective bit usage using masks will be sufficient to create a sufficiently reliable cryptographic primitive. Otherwise error correction methods, often based on classical communication theory, such as BCH decoding [36] or on newer cryptographic primitives such as learning with errors [21] are required. Many of these methods also require public helper data, but de-biasing methods are available to provably ensure no exploitable leakage is available to an adversary monitoring the communications [27].

9.3   Evaluation of an identity-based PUF for IoT applications

In order to create a secure, lightweight, privacy preserving network, such as that described in Reference 23, a reliable *weak* PUF is required such that minimal post-processing for error correction is required. A PUF suitable for such an application is now presented along with extensive experimental results.

9.3.1   FPGA-based PUF identifier

The compact PUF ID generator described by Gu *et al.* in Reference 18 comprises an array of *n* elementary 1-bit ID cells. The gate-level design of each 1-bit cell is schematically shown in Figure 9.3(a), where one 1-bit ID cell is capable of generating a single bit of a response ID and is manually placed and routed to ensure balanced routing. Two D flip-flops, are first reset by *CLEAR* and then activated by the rising edge of the *START* signal connected to their clock pins. As the flip-flops are coarse grain delay components, the propagation time of the two delay paths, T_0 and T_1, are

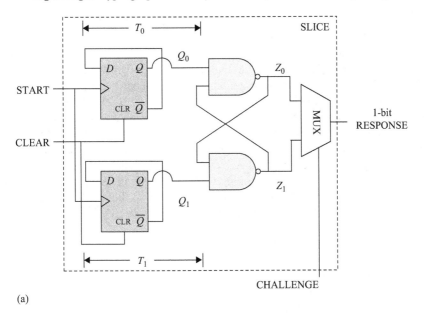

(a)

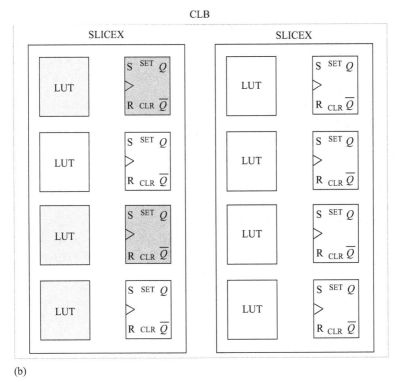

(b)

*Figure 9.3   PUF ID generator design. (a) 1-bit ID cell circuit design and
(b) Spartan-6 implementation*

different due to inherent manufacturing variability, hence they effectively race against each other. A cross-coupled 1-of-2 NAND arbiter forces the circuit into a metastable state, and determines which of the signals T_0 and T_1 arrives first. This sets its outputs, Z_0 and Z_1, to either $\{1,0\}$ or $\{0,1\}$. A $2 \rightarrow 1$ multiplexer then selects Z_0 or Z_1 as the 1-bit PUF ID generator response depending on the challenge bit. Hence, the proposed compact PUF design can be considered as a delay-based PUF. To implement the cell on a Spartan-6 FPGA, e.g., requires just three look-up tables (LUTs) and two flip-flops of a SLICEX as shown in Figure 9.3(b), and compactly fits into one FPGA slice. This allows all routing to be kept within the local slice switchblock without passing through the general interconnect, thereby increasing the reliability of the design due to the increased control over the timing paths. Note that the LUT required for the multiplexor is not required to be located within the same slice as the race condition has completed when the signals arrive at it, allowing a second ID cell to be located in the same slice if required.

To ensure the ID response bits are indeed a function of variability it is essential that all circuit placement and routing is symmetrically balanced, including the feedback paths of the arbiter. Otherwise, it is probable that the response bits will exhibit bias due to interconnect and layout mismatch. Suh *et al.* [41] suggest using FPGA hard-macros as a solution to meet tight design parameters, hence the proposed 1-bit compact PUF design can be created as a balanced hard-macro. This hard-macro cell helps to reduce the overall hardware resource consumption and minimize the routing congestion. Note, that the Xilinx Vivado [46] toolset depreciated the hard macro employed by their previous toolchain ISE [45] in favour of XDC macros. In this research work, an XDC macro is also created to prove the viability of the proposed PUF design for multiple generations of FPGA families.

9.3.1.1   Delay model

As mentioned previously *Weak* PUFs operate under different assumptions and are less complex than *strong* PUFs, thus utilising few or just a single challenge. In the extreme case *weak* PUFs generate just one response, e.g. SRAM PUF [19] and Butterfly PUF [25] and for many applications, their responses can be useful for key generation as an intrinsic key, in place of secure memory.

The proposed compact FPGA-based PUF ID generator design is a *weak* PUF and comprises an array of n elementary 1-bit cells. As shown in Figure 9.3(a), T_0 and T_1 represent the delay times from the input of the flip-flops to the input of the cross coupled NAND gates. The response may be zero when $\Delta T = T_0 - T_1$ is positive, otherwise the response will be one. Due to the compact architecture, a formula as shown in (9.4) can be used to directly express the relationship between each bit of the challenge and response. Hence, this architecture is not suitable for use in CRP based protocols where the CRPs are visible to an attacker.

$$T_0 - T_1 \underset{r=1}{\overset{r=0}{\gtrless}} 0 \qquad (9.4)$$

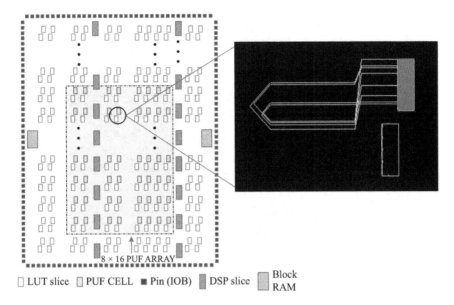

8 × 16 PUF ARRAY

☐ LUT slice   ☐ PUF CELL   ■ Pin (IOB)   ▌DSP slice   ▌Block RAM

Figure 9.4   128-bit PUF ID generator floor plan based on 1-bit single slice hard macro

The two flip-flops are coarse grain delay components, hence, the delay times of the signals from the input port D to the output port Q, T_0 and T_1, are physically different due to manufacturing variations. The cross coupled NAND gates, implemented as an arbiter here, determine the result Z_0 and Z_1 as $\{0, 1\}$ or $\{1, 0\}$. In theory, a cross-coupled NAND gate design can be also used as a standalone PUF design, however, the addition of the coarse grain flip-flops significantly extends the delay path, which means ΔT is increased to enhance the stability of the PUF circuit design. Additionally, the extra flip-flops can be situated in the same slice with the cross-coupled NAND gates without consuming any extra slice resource.

9.3.1.2 Implementation

As previously discussed, each single 1-bit PUF ID cell is implemented as a balanced hard-macro to ensure symmetrical placement and routing on an FPGA, which is shown in Figure 9.4. The routing is chosen so as to minimize the nominal delay difference between T_0 and T_1, and the arbiter feedback paths are balanced to achieve a fair arbiter. To construct the ID generator an array of n 1-bit ID cell hard-macros are instantiated in a Verilog netlist, which has single bit inputs *START*, *CLEAR*, an $[n-1:0]$ bus input for the challenge, and an $[n-1:0]$ bus output for the response ID. The floor plan location of each ID cell hard-macro, as shown in Figure 9.4, is set by declaring *LOC* constraints using the Xilinx unified constraints format (UCF) file format. This low level placement is required to prevent the tool chain optimising away the logic or randomly routing asymmetric paths.

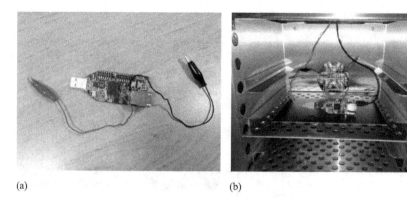

(a)   (b)

Figure 9.5   Experimental setup for 128-bit PUF ID generator. (a) Modified Avnet
LX9 Microboard and (b) heat chamber

9.3.1.3   Experimental evaluation

The proposed 128-bit PUF ID generator circuit was implemented and evaluated on ten different Avnet Spartan-6 LX9-2CSG324 Micro-Boards [44]. A simple 32-bit AMBA APB interface and serial port were added to test the designs via a PC running Matlab. One board was modified by breaking the core voltage supply line and soldering external wires across the de-coupling capacitor to conduct temperature and voltage experiments as shown in Figure 9.5. The experimental temperature range was tested from 20 °C to 70 °C, and the supply voltage of the FPGA's core was tested between 1.08 V and 1.32 V (0.02 V steps, 12 stages) using a regulated DC power supply, which is ±10% of the standard supply voltage of the FPGA's core voltage of 1.2 V.

9.3.1.4   Uniqueness

To investigate the uniqueness, responses were measured from the ten Spartan-6 devices ($k = 10$) implementing the proposed PUF ID generator using a randomly chosen challenge to generate 128-bit ID responses ($n = 128$). Figure 9.6 shows the average inter-chip HDs for the proposed PUF ID generator design. The distributions can be approximated by a Gaussian with the fitted line in red, and it can be seen that the average HD values for all cases are close to the center, i.e. 50%, indicating a high uniqueness. The uniqueness of the 1-bit single slice PUF ID generator design is 48%, which is close to the ideal value of 50% and can be attributed to the fact that the local routing when using a hard-macro based implementation, ensures balanced paths. The results validate that the proposed lightweight PUF ID generator circuit designs achieve a level of uniqueness suitable for IoT type applications.

9.3.1.5   Reliability

To measure the reliability of the PUF design, the experiments are repeated when changing the voltage between 1.08 V and 1.32 V in steps of 0.02 V, and the temperature range is changed from 20 °C to 70 °C. These ranges were chosen as they are the recommended operating ranges of the FPGA, and the embedded security monitors

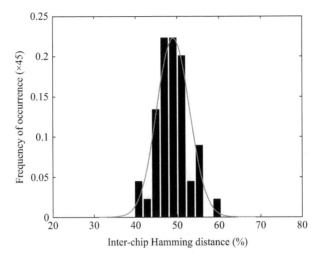

Figure 9.6   Uniqueness results

can be used to detect if an adversary tries to force the operating conditions outside these ranges. The results are shown in Figure 9.7, and it can be seen that the effect of temperature is less than that of voltage. Over all the operating points the PUF ID generator presents a high level of reliability at 94.27%. In this case, the localised routing allows for stable responses as the temperature changes, since both delay paths are affected equally as the temperature of the FPGA die increases.

9.3.1.6   Technology node

The Spartan-6 family of Xilinx FPGA devices manufactured at 45 nm process technology is just one of many FPGAs available from vendors such as Xilinx, Altera or Microsemi. As PUFs are based on manufacturing variability, there is no guarantee that what works at one process technology level will work on another, and designs may need to be adapted between different FPGAs to ensure that the localised routing is maintained. With this in mind the experiments were re-run on a number of Artix-7 Nexys4 boards [13]. This 7-series Xilinx FPGA family was manufactured at 28 nm, and is the same as that found in the FPGA layer of the Zynq SoC.

For these experiments, rather than implementing a 128-bit PUF circuit, a heat map was generated for the entire XC7A100T FPGA with a PUF cell placed in every single slice. In order to allow for control and communication circuitry (and allow a reasonable running time for implementation), this was done on a clock tile by clock tile basis. Reading each single bit response from every single cell 1 k times, the reliability of each slice is shown in Figure 9.8(a).[2] The distribution in Figure 9.8(b) shows that the majority of cells strongly return either {1, 0} hence majority voting can be utilised

---

[2]The blank areas are where auxiliary units such as the BRAM and DSP blocks reside.

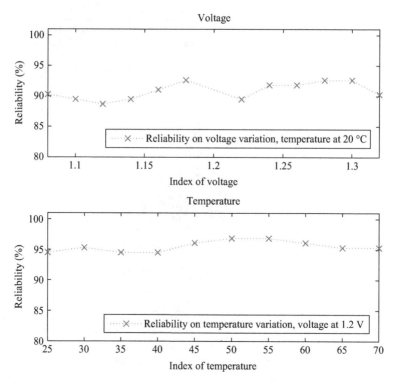

Figure 9.7   Reliability results: voltage and temperature variations of proposed 128 1-bit PUF ID cells

for these locations. Interestingly, a number of bits appear *truly* random returning {1, 0} with a probability of ≈0.5 which could have potential for random number generation, however further study is needed. There is also a small, but significant bias towards returning {1} rather than {0} hence de-biasing of the output is required prior to use in cryptographic operations.

This experiment was repeated on a further 9 FPGA boards, with the resultant bit probabilities calculated across all devices as shown in Figure 9.9. The Gaussian shaped distribution of Figure 9.9(b) shows that the PUF design operates as expected, as a given ID-bit implemented in a random slice will return a random response across different FPGAs.

9.3.1.7   Comparison of metrics for different PUFs

A large number of PUF proposals have been presented in the academic literature for both FPGA and ASIC targets. Table.9.1 compares the performance of some of the existing *weak* PUF designs on various devices. While designs across FPGA and ASIC implementations, or even different FPGA families or ASIC manufacturing processes, are not directly comparable, it does give an indication of the different resources

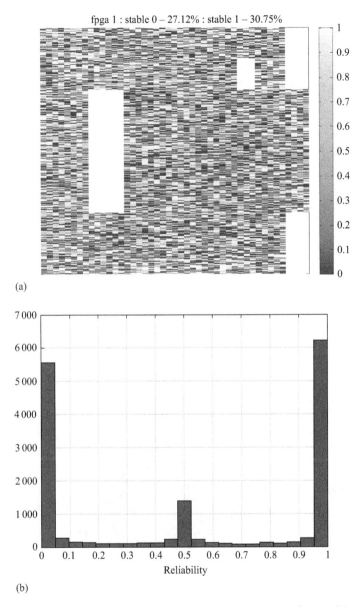

(a)

(b)

Figure 9.8   *Experimental results for single Artix-7 XC7A100T device. (a) Artix-7 slice reliability heatmap and (b) Artix-7 slice reliability distribution*

required. The reliability results of all PUF designs are given for varying temperature and/or voltage experiments wherever available.

Note the SRAM PUF proposed by Guajardo *et al.* [19] using SRAM memory cells, can only return a response on power-up. The Latch PUF proposed by Su *et al.*

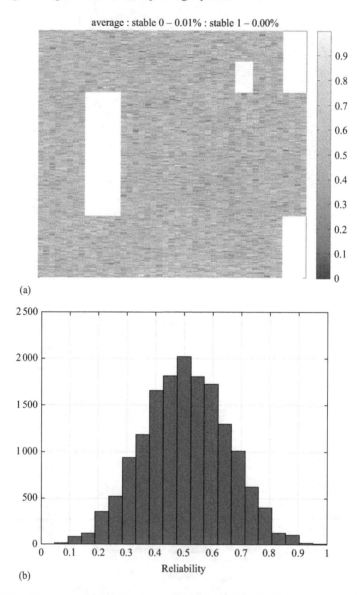

Figure 9.9   Experimental results for multiple Artix-7 XC7A100T devices. (a) Artix-7 slice reliability heatmap and (b) Artix-7 slice reliability distribution

[40] is very low power, however it is only reported on an ASIC. The flip-flop PUF proposed by Roel *et al.* [26], similar to SRAM, uses the power-up values of the flip-flops, however its randomness is limited and post-processing is required for improving the randomness. The Butterfly PUF proposed by Kumar *et al.* [25], which is also suitable for FPGA implementation as it can be implemented using basic logic gates,

Table 9.1   Comparison of metrics of different weak PUF designs

PUF design	Uniqueness	Reliability	Hardware	Bits	Resources
SRAM PUF [19]	49.97%	>88%[t]	FPGA	128	4600 SRAM
Latch PUF [40]	50.55%	96.96%	0.13 um CMOS	128	1 latch per cell
Latch PUF [48]	46%	>87%[t]	Spartan 3	128	2×128 slices
Flip-flop PUF [26]	$\approx$50%[pp]	>95%[pp]	Virtex 2	4096	4096 flip flops
Flip-flop PUF [43]	36%	>87%[t]	ASIC	1024	1024 flip flops
Buskeeper PUF [39]	49%	>80%[t] > 95%[v]	TSMC 65 nm	192	$1 GE^1$
Butterfly PUF [25]	$\approx$ 50%	>94%[t]	Virtex 5	64	130 slices
RO PUF [41]	46.15%	99.52%	Virtex 4	128	16×64 *array*^{2}
CRO PUF [29]	43.5%	>96%[t], $\approx$100%[v]	Spartan 3	127	64 slices for ROs
Proposed PUF	49.90%	94.53%	Artix 7	128	128 slices
Proposed PUF	48.52%	93.00%	Spartan 6	128	128 slices

[1]GE represents gate equivalent.
[2]16 $\times$ 64 array = 1024 ROs; each RO consisting of 5 inverters and 1 AND.
[t]Under temperature variation.
[v]Under supply voltage variation.
[pp]Required post-processing.

reported 94% reliability over temperature variations; however reliability over voltage changes is not considered. It consumes 130 slices of a Virtex-5 FPGA device for 64-bit response generation, which uses twice the hardware resources of the proposed 1-bit PUF ID generator design. The RO PUF proposed by Suh *et al.* [41] and the CRO PUF proposed by Maiti *et al.* [29] have been implemented on different FPGAs, e.g. Virtex-4 and Spartan-3. The hardware resource consumption is at least 384 slices for a 64-bit response. It can be seen that the proposed PUF design is the most lightweight FPGA-based Weak PUF ID generator design reported to date.

9.4   Discussion

As the number of connected IoT devices grows exponentially, traditional crypto-graphic key management solutions for securing them become problematic and are often too computationally intense for such lightweight devices. In this chapter, a broad overview of a promising security primitive for IoT, known as PUFs, has been presented. The differences between *strong* and *weak* PUFs are given, as well as their suitability for integration into lightweight applications for securing IoT devices. The factors that need to be considered in the design of an effective PUF are discussed and can be summarised as follows:

- Is a *strong* or *weak* PUF needed for the target application?
- Select a suitable design approach (delay-based, memory-based or RO-based) based on time/area/entropy trade-offs
- Evaluate PUF design on target device (rather than simulation)
- Comprehensively test design for uniqueness, reliability, robustness, uniformity, bit-aliasing and entropy
- Consider attack scenarios and need for countermeasures against modelling and SCA attacks
- Consider end-to-end design aspects of PUF-based application, e.g. number of CRPs needed, acceptable error rate, use of reverse fuzzy extractors
- Select an appropriate error correction method

A lightweight identity-based PUF design (*weak* PUF) suitable for IoT applications is also presented and evaluated on two different FPGA technologies (45 nm Spartan-6 and 28 nm Artix-7 devices). In comparison to other proposed PUF design approaches, this is the most lightweight FPGA-based *weak* PUF design. Moreover, the empirical results show the effectiveness of the proposed design in terms of the important metrics of uniqueness and reliability.

References

[1]   Intrinsic-id. www.intrinsic-id.com. Last Accessed 11.04.2016.
[2]   Sirona. www.sirona.io. Last Accessed 11.04.2016.
[3]   Verayo. www.verayo.com. Last Accessed 11.04.2016.
[4]   Frederik Armknecht, Roel Maes, Ahmad-Reza Sadeghi, François-Xavier Standaert, and Christian Wachsmann. A formalization of the security features of physical functions. In *32nd IEEE Symposium on Security and Privacy, S&P 2011, 22–25 May 2011, Berkeley, California, USA*, pp. 397–412. IEEE Computer Society, Oakland, 2011.
[5]   Aydin Aysu, Ege Gulcan, Daisuke Moriyama, Patrick Schaumont, and Moti Yung. End-to-end design of a puf-based privacy preserving authentication protocol. Tim Güneysu and Helena Handschuh, editors. *Cryptographic Hardware and Embedded Systems—CHES 2015 – 17th International Workshop, Saint-Malo, France, September 13–16, 2015, Proceedings, Lecture Notes in Computer Science*, vol. 9293, Springer, Saint-Malo, 2015, pp. 556–576.
[6]   Georg T. Becker. The gap between promise and reality: On the insecurity of XOR arbiter pufs. Tim Güneysu and Helena Handschuh, editors. *Cryptographic Hardware and Embedded Systems—CHES 2015 – 17th International Workshop, Saint-Malo, France, September 13–16, 2015, Proceedings, Lecture Notes in Computer Science*, vol. 9293, Springer, Saint-Malo, 2015, pp. 535–555.

[7]   Georg T. Becker and Raghavan Kumar. Active and passive side-channel attacks on delay based PUF designs. *IACR Cryptology ePrint Archive*, 2014:287, 2014.

[8]   Cisco. Internet of things IoT. http://www.cisco.com/web/solutions/trends/ iot/portfolio.html. Accessed: 23-07-2015.

[9]   Cryptovia. ECDSA for AVR CPU. http://www.cryptovia.com/AVR_ ECDSA.html. Accessed: 27-05-2015.

[10]   Jeroen Delvaux, Dawu Gu, Roel Peeters, and Ingrid Verbauwhede. Secure lightweight entity authentication with strong pufs: Mission impossible II. *IACR Cryptology ePrint Archive*, 2014:977, 2014.

[11]   Jeroen Delvaux, Dawu Gu, Roel Peeters, and Ingrid Verbauwhede. A survey on lightweight entity authentication with strong PUFs. Cryptology ePrint Archive, Report 2014/977, 2014. http://eprint.iacr.org/.

[12]   Jeroen Delvaux, Dawu Gu, Dries Schellekens, and Ingrid Verbauwhede. Secure lightweight entity authentication with strong pufs: Mission impossible? In Lejla Batina and Matthew Robshaw, editors, *Cryptographic Hardware and Embedded Systems – CHES 2014 – 16th International Workshop, Busan, South Korea, September 23–26, 2014. Proceedings, Lecture Notes in Computer Science*, vol. 8731, Springer, Busan, 2014, pp. 451–475.

[13]   Digilent. Nexys4 Artix-7 FPGA Board. http://digilentinc.com/ Products/Detail.cfm?NavPath=2,400,1184&Prod=NEXYS4. Accessed: 03-11-2015.

[14]   Yevgeniy Dodis, Rafail Ostrovsky, Leonid Reyzin, and Adam Smith. Fuzzy extractors: How to generate strong keys from biometrics and other noisy data. *SIAM J. Comput.*, 38(1):97–139, 2008.

[15]   Maria Ganzha, Leszek A. Maciaszek, and Marcin Paprzycki, editors. *Proceedings of the 2014 Federated Conference on Computer Science and Information Systems*, Warsaw, Poland, September 7–10, 2014, 2014.

[16]   Blaise Gassend, Dwaine Clarke, Marten Van Dijk, and Srinivas Devadas. Silicon physical random functions. In *Proceedings of the 9th ACM Conference on Computer and Communications Security, CCS 2002*, Washington, DC, USA, November 18–22, 2002, 2002.

[17]   Chongyan Gu, Neil Hanley, Robert Hesselbarth, Martin Hutle, and Gavin McWilliams. "Sparks Deliverable 4.2 – PUF Enhanced Smart Meter Hardware Architecture and an Authentication/Key Management Deployment Architecture (interim)", 2015.

[18]   Chongyan Gu, Julian Murphy, and Máire O'Neill. A unique and robust single slice FPGA identification generator. In *Proc. IEEE International Symposium on Circuits and Systems (ISCAS'14)*, Melbourne, Australia, June 2014, pp. 1223–1226.

[19]   Jorge Guajardo, Sandeep S. Kumar, Geert Jan Schrijen, and Pim Tuyls. FPGA intrinsic pufs and their use for IP protection. In Pascal Paillier and Ingrid Verbauwhede, editors, *Cryptographic Hardware and Embedded Systems – CHES 2007, Ninth International Workshop, Vienna, Austria, September 10–13, 2007, Proceedings, Lecture Notes in Computer Science*, vol. 4727, Springer, Vienna, 2007, pp. 63–80.

[20]   Tim Güneysu and Helena Handschuh, editors. *Cryptographic Hardware and Embedded Systems – CHES 2015 – 17th International Workshop, Saint-Malo, France, September 13–16, 2015, Proceedings, Lecture Notes in Computer Science*, vol. 9293, Springer, Saint-Malo, 2015.

[21]   Charles Herder, Ling Ren, Marten van Dijk, Meng-Day (Mandel) Yu, and Srinivas Devadas. A stateless cryptographically-secure physical unclonable function. *IACR Cryptology ePrint Archive*, 2015:798, 2015.

[22]   Anthony Van Herrewege, Stefan Katzenbeisser, Roel Maes, *et al.*, Reverse fuzzy extractors: Enabling lightweight mutual authentication for puf-enabled rfids. In Angelos D. Keromytis, editor, *Financial Cryptography and Data Security – 16th International Conference, FC 2012, Kralendijk, Bonaire, February 27–March 2, 2012, Revised Selected Papers, Lecture Notes in Computer Science*, vol. 7397, Springer, Kralendijk, 2012, pp. 374–389.

[23]   C. Huth, J. Zibuschka, P. Duplys, and T. Guneysu. Securing systems on the internet of things via physical properties of devices and communications. In *Systems Conference (SysCon), 2015 Ninth Annual IEEE International*, April 2015, pp. 8–13.

[24]   Ilya Kizhvatov. Side channel analysis of AVR XMEGA crypto engine. In Dimitrios N. Serpanos and Wayne Wolf, editors, *Proceedings of the Fourth Workshop on Embedded Systems Security, WESS 2009*, Grenoble, France, October 15, 2009. ACM, 2009.

[25]   Sandeep S Kumar, Jorge Guajardo, Roel Maes, G-J Schrijen, and Pim Tuyls. The butterfly PUF protecting IP on every FPGA. In *Proc. IEEE International Symposium on Hardware-Oriented Security and Trust (HOST'08)*, 2008, pp. 67–70.

[26]   Roel Maes, Pim Tuyls, and Ingrid Verbauwhede. Intrinsic PUFs from flip-flops on reconfigurable devices. In *Proc. Third Benelux Workshop on Information and System Security (WISSec 2008)*, p. 17, Eindhoven, the Netherlands, 2008.

[27]   Roel Maes, Vincent van der Leest, Erik van der Sluis, and Frans Willems. Secure key generation from biased pufs. Tim Güneysu and Helena Handschuh, editors. *Cryptographic Hardware and Embedded Systems – CHES 2015 – 17th International Workshop, Saint-Malo, France, September 13–16, 2015, Proceedings, Lecture Notes in Computer Science*, vol. 9293, Springer, Saint-Malo, 2015, pp. 517–534.

[28]   Abhranil Maiti, Vikash Gunreddy, and Patrick Schaumont. A systematic method to evaluate and compare the performance of physical unclonable functions. *IACR Cryptology ePrint Archive*, 2011:657, 2011.

[29]   Abhranil Maiti and Patrick Schaumont. Improving the quality of a physical unclonable function using configurable ring oscillators. In *Proc. 19th IEEE International Conference on Field Programmable Logic and Applications (FPL'09)*, Czech Republic, September 2009, pp. 703–707.

[30]   Abhranil Maiti and Patrick Schaumont. A novel microprocessor-intrinsic physical unclonable function. In *Proc. 22nd International Conference on Field Programmable Logic and Applications (FPL'12)*, pp. 380–387, Oslo, Norway, August 2012.

[31]   Jonathan M. McCune, Boris Balacheff, Adrian Perrig, Ahmad-Reza Sadeghi, Angela Sasse, and Yolanta Beres, editors. *Trust and Trustworthy Computing – Fourth International Conference, TRUST 2011, Pittsburgh, PA, USA, June 22–24, 2011. Proceedings, Lecture Notes in Computer Science*, vol. 6740. Pittsburgh, Springer, 2011.

[32]   Dominik Merli, Johann Heyszl, Benedikt Heinz, Dieter Schuster, Frederic Stumpf, and Georg Sigl. Localized electromagnetic analysis of RO pufs. In *2013 IEEE International Symposium on Hardware-Oriented Security and Trust, HOST 2013*, Austin, TX, USA, June 2–3, 2013, pp. 19–24. IEEE Computer Society, Austin, 2013.

[33]   David Naccache and Patrice Fremanteau. "Unforgeable identification device, identification device reader and method of identification", July 1995. US Patent 5,434,917.

[34]   David Oswald, Bastian Richter, and Christof Paar. Side-channel attacks on the yubikey 2 one-time password generator. In Salvatore J. Stolfo, Angelos Stavrou, and Charles V. Wright, editors, *Research in Attacks, Intrusions, and Defenses – 16th International Symposium, RAID 2013*, Rodney Bay, St. Lucia, October 23–25, 2013. *Proceedings, Lecture Notes in Computer Science*, vol. 8145, Springer, Rodney Bay, 2013, pp. 204–222.

[35]   Ravikanth Pappu, Ben Recht, Jason Taylor, and Neil Gershenfeld. Physical one-way functions. *Science* 297(5589):2026–2030, 2002.

[36]   Emmanuel Prouff and Patrick Schaumont, editors. *Cryptographic Hardware and Embedded Systems – CHES 2012 – 14th International Workshop, Leuven, Belgium, September 9–12, 2012. Proceedings, Lecture Notes in Computer Science*, vol. 7428, Springer, Leuven, 2012.

[37]   Ulrich Rührmair and Jan Sölter. PUF modeling attacks: An introduction and overview. In *Proc. Design, Automation & Test in Europe Conference & Exhibition (DATE'14)*, Dresden, Germany, March 2014, pp. 1–6.

[38]   Ulrich Rührmair, Jan Sölter, Frank Sehnke, *et al.* PUF modeling attacks on simulated and silicon data. *IEEE Trans. Inf. Forensics Security*, 8(11):1876–1891, 2013.

[39]   Peter Simons, Erik van der Sluis, and Vincent van der Leest. Buskeeper PUFs, a promising alternative to D flip-flop PUFs. In *Proc. IEEE International Symposium on Hardware-Oriented Security and Trust (HOST'12)*, San Francisco, CA, USA, June 2012, pp. 7–12.

[40]   Ying Su, Jeremy Holleman, and Brian P Otis. A digital 1.6 pj/bit chip identification circuit using process variations. *IEEE J. Solid-State Circuits*, 43:69–77, 2008.

[41]   G. Edward Suh and Srinivas Devadas. Physical unclonable functions for device authentication and secret key generation. In *Proceedings of the 44th Design Automation Conference, DAC 2007, San Diego, CA, USA, June 4–8, 2007*, San Diego, IEEE, 2007, pp. 9–14.

[42]   Shahin Tajik, Enrico Dietz, Sven Frohmann, *et al.* Physical characterization of arbiter pufs. In *Cryptographic Hardware and Embedded Systems – CHES 2014*, Springer, Busan, South Korea, 2014, pp. 493–509.

[43]   Vincent van der Leest, Geert-Jan Schrijen, Helena Handschuh, and Pim Tuyls. Hardware intrinsic security from D flip-flops. In *Proceedings of the Fifth ACM workshop on Scalable Trusted Computing*, ACM, Chicago, IL, USA, 2010, pp. 53–62.

[44]   Xilinx. Avnet spartan-6 LX9 microboard. http://www.xilinx.com/products/boards-and-kits/1-3i2dfk.html. Accessed: 23-07-2015.

[45]   Xilinx. ISE design suite. Accessed: 23-07-2015.

[46]   Xilinx. Vivado design suite. Accessed: 23-07-2015.

[47]   Xilinx. Zynq-7000 All Programmable SoC. http://www.xilinx.com/products/silicon-devices/soc/zynq-7000.html. Accessed: 27-05-2015.

[48]   Dai Yamamoto, Kazuo Sakiyama, Mitsugu Iwamoto, *et al.*   Uniqueness enhancement of PUF responses based on the locations of random outputting RS latches. In Bart Preneel and Tsuyoshi Takagi, editors, *Cryptographic Hardware and Embedded Systems*, *Lecture Notes in Computer Science*, vol. 6917, Springer, Berlin, 2011, pp. 390–406.

Chapter 10

A reputation model for the Internet of Things

Benjamin Aziz, Paul Fremantle, and Alvaro Arenas

Summary

The MQ Telemetry Transport (MQTT) protocol has emerged over the past decade as a key protocol for a number of low power and lightweight communication scenarios including machine-to-machine and the Internet of Things. In this chapter we develop a utility-based reputation model for MQTT, where we can assign a reputation score to participants in a network based on monitoring their behaviour. We mathematically define the reputation model using utility functions on participants based on the expected and perceived behaviour of MQTT clients and servers. We define an architecture for this model, and discuss how this architecture can be implemented using existing MQTT open source tools, and we demonstrate how experimental results obtained from simulating the architecture compare with the expected outcome of the theoretical reputation model.

10.1 Introduction

The Internet of Things (IoT) is an area where there is significant growth: both in the number of devices deployed and in the scenarios in which devices are being used. One of the challenges for the Internet of Things is supporting network protocols which utilise less energy, lower bandwidth, and support smaller footprint devices. One such protocol is the MQ Telemetry Transport (MQTT) protocol [1], which was originally designed to support remote monitoring and Supervisory Control And Data Acquisition (SCADA) scenarios but has become popular for the IoT.

Another challenge with IoT networks is that small devices may not perform as well as needed due to a number of factors including: network outages or poor network performance due to the use of 2G or other low bandwidth networks, power outages for devices powered by batteries, deliberate vandalism or environmental damage for devices placed in public areas, and many other such challenges. Therefore we identified that a reputation model for devices connecting by MQTT would be a useful construct to express consumers' (applications') trust in the behaviour and performance of these devices as well as measure the level of performance of the server aggregating data from such devices according to some predefined Service Level

Agreement (SLA). In addition, we implemented the reputation model to demonstrate that it could be used in real MQTT networks.

Our model of reputation is based on the notion of a *utility function*, which formally expresses the consumer's level of satisfaction related to various issues of interest against which the reputation of some entity is measured. In the case of MQTT networks, one notable such issue is the Quality of Service (QoS) with regards to the delivery of messages; whether messages are delivered exactly once, more than once or at most once to their consumers. The model, inspired by previous works [2, 3], is general enough to be capable of defining the reputation of client devices and servers at various levels of abstraction based on their level of performance in relation to the delivery of messages issue of interest.

The chapter starts with an overview of the MQTT protocol (Section 10.3). From this, we then mathematically define the reputation model for MQTT clients and server (Section 10.4), based on their ability to keep to the requirements of the protocol. We then outline a system architecture (Section 10.5) for monitoring the MQTT protocol and thereby being able to calculate the reputation by observing the behaviour of MQTT clients and server in a real network. We show how this system was implemented and we demonstrate the results of this implementation (Section 10.6). Finally, we look at related work (Section 10.7) and conclude the chapter outlining areas for further research (Section 10.8).

10.2   Overview of the concept of reputation

Reputation is a general concept widely used in all aspects of knowledge ranging from humanities, arts and social sciences to digital sciences. It is a concept closely related to trust and it is defined by the Merriam-Webster dictionary [4] as the "overall quality or character as seen or judged by people in general". In fact, reputation is often seen as one measure by which trust or distrust can be built based on good or bad past experiences and observations (direct trust) [5] or based on collected referral information (indirect trust) [6]. In recent years, the concept of reputation has shown itself to be useful in many areas of research in computer science, particularly in the context of distributed and collaborative systems, where interesting issues of trust and security manifest themselves. Therefore, one encounters several definitions, models and systems of reputation in distributed computing research, of which we attempt to touch upon a few briefly in the upcoming sections.

Trust and reputation systems have been recognised as playing an important role in decision making in the Internet world [5, 7]. Customers and sellers might trust themselves and the services they are offering. According to Reference 8, trust refers to the subjective probability by which an individual expects that another performs a given action on which its welfare depends. Grandison and Soloman [7] refer to trust as being the firm belief in the competence of an entity to act dependably, securely and reliably within a specified context. A difference is noted in Reference 5 between *reliability trust* as a subjective probability and the *decision trust* as being the extent in which one party is willing to depend on something or somebody in a given situation with a feeling of relative security, even though negative consequences are possible.

A reputation management system should address the following challenges [9]: (1) how an agent rates the correspondent based on the past interaction history, (2) how an agent finds the right witnesses in order to select the referral agents and (3) how the agent systematically incorporates the testimonies of those witnesses.

From the theoretical point of view, reputation is formalised in a game-theoretical framework, where agents are continuously playing the same game with incomplete information. The main concern is when and whether a long-lived player can take advantage of a small probability of a certain type or reputation to effectively commit him to playing as if he were that type [10]. Game theory usual helps one to demonstrate that it is worth to consider reputation information when analysing the outcome of some competing situations with incomplete information. We direct the reader to the work of Reference 11 that theoretically analyses the sensitive options a reputation system designer might have.

Reputation-based systems are mainly used in electronic markets (e.g. eBay) as a way of assessing the participants. In such environments they proved to be effective as the number of participants is very large and the system is running a sufficient amount of time [12]. As service-oriented computing environments become the basis for the future ubiquitous and pervasive computing, with highly dynamic, volatile and independent service consumers and providers, reputation management will become mandatory for enhancing these systems with trust.

Reputation could be based on direct experience or indirect information. In the direct approach, the simplest method is to compute the reputation of individuals based on aggregating the received feedback on transactions and convert this feedback into a reputation measure. This approach is widely used in Internet e-commerce sites, where users rate the providers. Internet sites (e.g. eBay and Amazon) mainly use summation-based centralised reputation systems, based on counting all votes or grades an entity receives. Their big advantage is the simplicity of the reputation scheme. This makes the reputation value to be easily understood by the participants and allows a direct conversion between reputation assessment and trust.

Zacharia and Maes [13] were among the first ones to build more sophisticated reputation-based systems to overcome existing trust problems in e-commerce on-line applications. They proposed the SPORAS system, based only on direct transaction ratings between users. Users rate each other after a transaction with continuous values from 0.1 to 1. The ratings of a user are aggregated in a recursive fashion, obtaining a reputation value that scales from 0 to 3 000 and a reputation deviation to assess the reliability of the reputation value. SPORAS was evaluated to perform better than the eBay and Amazon approaches.

In the indirect approach, building trust is not only based on the past interactions between entities, but also considers the social network the entity belongs to and the referrals the entity can obtain using the social network. This approach is widely considered in multi-agent research, where e.g. Reference 14 defines the concepts of agent communities and social networks. The members of an online community provide services and referrals for services to each other. A participant in a social network has reputation for both *expertise* (providing good services) and *sociability* (providing good referrals). This theoretical framework is conceptually valid also for P2P service-oriented systems. Items under study concern the way the reputation is represented, how referral information is aggregated, which learning model is used.

For the rest of the chapter, we mainly follow the direct reputation approach based on the rating of the levels of quality of service related to the delivery of messages according to some expectation defined based on utility functions.

10.3   MQTT

MQTT [15] is described as a lightweight broker-based publish/subscribe messaging protocol that was designed to allow devices with small processing power and storage, such as those which the IoT is composed of, to communicate over low-bandwidth and unreliable networks. The publish/subscribe message pattern [16], on which MQTT is based, provides for one-to-many message distribution with three varieties of delivery semantics, based on the level of QoS expected from the protocol. In the "at most once" case, messages are delivered with the best effort of the underlying communication infrastructure, which is usually IP-based, therefore there is no guarantee that the message will arrive. This protocol is termed the $QoS = 0$ protocol. In the second case of "at least once" semantics, certain mechanisms are incorporated to allow for message duplication. Despite the guarantee of delivering the message, there is no guarantee that duplicates will be suppressed. This protocol is also known as the $QoS = 1$ protocol. Finally, for the last case of "exactly once" delivery semantics, also known as the $QoS = 2$ protocol, the published message is guaranteed to arrive only once at the subscribers. The protocol also defines message structures needed in communications between *clients*, i.e. end-devices responsible for generating data from their domain (the data source) and *servers*, which are the system components responsible for collating source data from clients/end-devices and distributing these data to interested subscribers. Servers are often also referred to as *brokers*, as they intermediate between the data publishers and subscribers.

10.4   A reputation model for MQTT

We show in this section how the model of reputation defined for business processes in References 17, 18 can be adapted, with minimum changes, to the MQTT protocol to obtain the reputation of client devices and the server.

10.4.1   Monitoring events

Central to the model defined by Aziz and Hamilton [17, 18] was the notion of an *event*, which is a signal produced by an independent *monitor system*, which is monitoring the interactions occurring between the different entities in the monitored environment, in this case the client and server entities participating in the MQTT protocol. An event is defined as follows:

$$| \ Event : TimeStamp \times Ag \times Msg \times Id \times \mathbb{N}$$

where *TimeStamp* is the timestamp of the event generated by the monitor system issuing it, *Ag* is the identity of the agent (client device or server) to whom the event is related, *Msg* is the specific message of interest (e.g. *Publish* and *Pubrel* messages), *Id* is an identity value of the protocol instance and finally, $\mathbb{N}$ is a natural number representing the number of times the message *Msg* has been monitored, i.e. was sent.

For example, the following event, issued at monitor system's local time:

$$ev_{ex1} = (12:09:52, temp\_sensor, Publish, 1234, 2)$$

denotes that the *temp_sensor* device has been monitored, within the instance number 1234 of the protocol, to have sent twice the *Publish* message to the server responsible for collecting environment temperature data. On the other hand, the following event issued at local time $12:19:02$:

$$ev_{ex2} = (12:19:02, temp\_server, Publish, 1234, 1)$$

denotes that the server responsible for the environment temperature, *temp_server*, has been monitored, within the same instance number 1234 of the protocol, to have published only once the specific message *Publish* to the subscribers of the temperature topic. In both these examples, the assumption is that the monitor system is capable of detecting that the protocol instance being monitored has terminated before it issues any events related to that instance. Although theoretically this is impossible due to the halting problem, in practical terms, the monitor system can assume the protocol to have terminated after some reasonable amount of time has elapsed since the last protocol message.

The monitor generates events in the above form, which are used by a *reputation engine* to determine the reputation values for client devices and servers in an MQTT-based environment. The reputation engine will then use a *utility function* pre-supplied to the engine by subscribers to determine the level of satisfaction of a subscriber with regards to the results reported within an event:

$$\begin{array}{|l} utility : Event \times SLA \rightarrow [0, 1] \\ \hline \forall\, (t, a, m, i, n) \in Event, sla \in SLA \ \bullet \ utility((t, a, m, i, n), sla) = r \in \mathbb{R} \end{array}$$

This utility function will consider a SLA, defined as follows:

$$| \ \ SLA : Ag \times Top \times Iss \rightarrow \mathbb{N}^0$$

Here the SLA considers an issue of interest to the subscriber, *Iss*, which will be in our case the QoS level value fixed to one of 0, 1 or 2, expected from a particular agent *Ag* in relation to a specific topic *Top*. The outcome of the utility function is a real number *r* representing the satisfaction level of the subscriber in terms of both the SLA and the real values reported by events.

For example, consider the following SLA instance

$$sla = ((temp\_server, temperature, QoS), 2)$$

then given the event ev_{ex2}, the utility function could return the following value:

$$utility(t, temp\_server, Publish, 1234, 1, ((temp\_server, temperature, QoS), 2)) = 1$$

This indicates that the subscriber's requirements have been fulfiled, as indicated by their SLA ($r = 1$), with the results reported by the event ev_{ex2}. On the other hand, considering the same SLA, the utility function might return:

$$utility(t, temp\_server, Publish, 1234, 0, ((temp\_server, temperature, QoS), 2)) = 0$$

to show that the subscriber has a satisfaction value of 0 since the number of times the message was delivered to the subscriber is lower (i.e. 0) than what its QoS level is defined in the SLA (i.e. 2), therefore breaching the exactly once delivery semantics to the subscriber principle in MQTT.

Since the number of times a message is delivered will either confirm or not to the level of QoS expected by the subscriber, in all of the above cases, the score given will reflect either total satisfaction (i.e. 1) or total dissatisfaction (i.e. 0).

10.4.2   Reputation models

After introducing the main notions of an event and a utility function, we can now define models of reputation for the clients (e.g. sensor devices) and the MQTT server (broker) that aggregates the messages from the clients before publishing them to the subscribers. The subscribers are assumed to be the business applications or data consumers, and we do not include them in the reputation model. The MQTT standard does not prohibit a client from acting as both a device (i.e. source of data) and a subscriber (i.e. consumer of data). However, in our case, we only measure the reputation of the "source of data" clients.

10.4.2.1   The server reputation model

The first reputation model reflects the behaviour of MQTT servers. Given a set of events, *Event*, captured by the monitor system and relevant to the server for whom the reputation is being calculated, then we can define the server's reputation function computed at a particular point in time and parameterised by a specific SLA as follows:

$$
\begin{array}{l}
\boxed{=[Srv, SLA, TimeStamp]=} \\
s\_rep\_sla : Srv \times SLA \times TimeStamp \to [0,1] \\
\hline
\forall\, eset_s : \wp(Event) \bullet \\
s\_rep\_sla(s, sla, t) = \dfrac{\sum\limits_{ev \in eset_s.snd(ev)=fst(sla)=s \,\wedge\, id\_top(ev,sla)} \varphi(t,te)\, utility(ev,sla)}{\#eset_s}
\end{array}
$$

where #s denotes the cardinality of a set s and $\varphi(t, te)$ is a time discount function that puts more importance (emphasis) on events registered closer in time to the moment of computing the reputation. One definition of $\varphi(t, te)$ could be the time discount function defined by Huynh *et al.* [19], which we redefine here as $\varphi(t, te) = e^{-\frac{t-te}{\lambda}}$, where t is the current time at which the reputation is calculated, te is the timestamp of the event being considered and λ is *recency scaling factor* used to adjust the value of the function to a scale required by the application. After this, the server reputation function, *s_rep_sla*, is defined as the weighted average of the utilities obtained from all the generated events with respect to some SLA.

The above definition aggregates the set of all relevant events, i.e. the events that first have the same server name as that appearing in the SLA and second that are on an instance of the protocol related to the topic of the SLA. The first condition is checked using the two operators *fst* and *snd*, which will return the first and second elements of a tuple, whereas the second condition is checked using the predicate *id_top(ev, sla)*, which returns a True outcome if and only if the identity number of an instance of a protocol captured by *ev* corresponds to the topic value mentioned in the SLA *sla*. Considering the example events of the previous section, we would have the following calculation of *id_top(ev, sla)*:

id_top(12 : 09 : 52, *temp_server*, *Publish*, 1234, 1, ((*temp_server*, *temperature*, *QoS*), 2)) = *True*

The above definition calculates the sum of the time-discounted utility function values, with respect to the given SLA and the events gathered, and average these over the total number of events gathered (#$eset_s$) in any one instance when this reputation value is calculated.

Based on the definition of *s_rep_sla*, we next aggregate the reputation of a server across every SLA that binds that server to its subscribers:

$$
\begin{array}{|l|}
\hline
\rule{0pt}{0pt}\text{[}Srv, TimeStamp\text{]} \\
\quad s\_rep : Srv \times TimeStamp \rightarrow [0, 1] \\
\hline
\quad \forall\, slaset_s : \wp(SLA) \bullet s\_rep(s, t) = \dfrac{\displaystyle\sum_{sla\, \in\, slaset_s, fst(sla)\, =\, s} s\_rep\_sla(s, sla, t)}{\#slaset_s} \\
\hline
\end{array}
$$

which provides a more general indication of how well a server s behaves in relegation to a number of subscribers. This reputation is again calculated in a particular point in time, t, however it is straightforward to further generalise this reputation function over some time range, between t and t'.

10.4.2.2   The client device reputation model

After introducing the reputation model of the server, we define here the client's reputation model. Like the server, a client might also be implementing the QoS correctly, but it requires multiple reconnections, duplicate messages, etc., while the server does not. For instance, if the devices are not sending PINGs or responding to them, or this is delayed, it might indicate that a problem is more likely to occur in the future. Similarly, if the device needs to send multiple duplicate messages or needs to be sent duplicate messages, it also might indicate possible failure in the future. Thus, the reputation model for a client may be based on either the "Keep Alive/PING" case or the "Client's Retransmission Procedure" case. However, we start with the definition of an overall reputation model that generalises these two cases.

Given a set of events, *Event*, captured by the monitor system relevant to some client, then we define the client's reputation function computed at a particular point in time in a specific process (Keep Alive/PING procedure or retransmission procedure) and parameterised by a specific SLA as follows:

$$
\begin{array}{l}
\text{[\textit{Client, SLA, TimeStamp, Procedure}]} \\[4pt]
\hline
c\_rep\_sla\_p : Client \times SLA \times TimeStamp \times Procedure \rightarrow [0,1] \\[4pt]
\hline
\forall\, pset_s : \wp(Event) \bullet c\_rep\_sla\_p(c, sla, t, p) = \\[4pt]
\dfrac{\displaystyle\sum_{ev \in pset_s.snd(ev)\,=fst(sla)\,=\,c\,\wedge\,id\_top(ev,sla)} \varphi(t,te)\; utility(ev,sla)}{\#pset_s}
\end{array}
$$

This definition gathers the set of all related events, i.e. the events that first have the same client name as that appearing in the SLA and second that are on an instance of the protocol related to the topic of the SLA. The definition is parameterised by the client, an SLA, a timestamp and the specific procedure (e.g. Keep Alive/PING or retransmission). The SLA represents what the expectation is, from the server's point of view, of the client's behaviour in the context of the specific procedure. Similar to the case of *s_rep_sla*, a utility function is applied to measure the satisfaction of the server, in a time-discounted manner, in relation to the client's behaviour and this is then averaged over the total number of events captured in a specific instance of time.

For example, consider the case of the Keep Alive/PING procedure, then *c_rep_sla_ka* is defined as the time-discounted average of the utilities obtained from all generated events with respect to the Keep Alive/PING procedure.

$$
\begin{array}{l}
\text{[\textit{Client, SLA, TimeStamp, KeepAlive}]} \\[4pt]
\hline
c\_rep\_sla\_ka : Client \times SLA \times TimeStamp \times KeepAlive \rightarrow [0,1] \\[4pt]
\hline
\forall\, pset_s : \wp(Event) \bullet c\_rep\_sla\_ka(c, sla, t, ka) = \\[4pt]
\dfrac{\displaystyle\sum_{ev \in kapingset_s.snd(ev)\,=fst(sla)\,=c\,\wedge\,id\_top(ev,sla)} \varphi(t,te)\; utility(ev,sla)}{\#kapingset_s}
\end{array}
$$

In this procedure, the client sends a Pingreq message within each KeepAlive time period, then the receiver answer with a Pingresp message when it receives a Pingreq message from the gateway to which it is connected. Clients should use KeepAlive timer to observe the liveliness of the gateway to check whether they are connected to broker. If a client does not receive a Pingresp from the gateway even after multiple retransmissions of the Pingresq message, it fails to connect with gateway during the Keep Alive period.

Hence, for the above example, using *id_top* to show a set of related events *ev* corresponds to the topic value mentioned in the SLA, *sla*, we would have that:

$$id\_top(12 : 09 : 52, client, Pingreq, False, False, 1234, 1, ((client, temperature, QoS), 0)) = True$$

The event $ev_{kaping} = (12 : 09 : 52, client, Pingreq, False, False, 1234, 1)$ generated by the monitor could reflect a client device that has sent once the Pingreq message to connect to the gateway within the instance number 1234 of the protocol during the Keep Alive period. Then, given the SLA instance $sla = ((client, temperature, QoS), 0)$, the client should deliver this Pingreq message in relation to a specific topic (in this case *temperature*) at most once within each KeepAlive time period, but there is no guarantee the message will arrive.

From the definition of *c_rep_sla_p*, we generate a more general reputation for some client in a particular point in time *t* within a period, *Period*, as follows:

$$
\begin{array}{l}
=[Client, SLA, TimeStamp]=\\
\quad c\_rep\_sla : Client \times SLA \times TimeStamp \rightarrow [0, 1]\\
\quad\quad \forall\, periodset_s : \wp(Period) \bullet c\_rep_sla(c, sla, t) =\\
\quad\quad\quad \dfrac{\displaystyle\sum_{ev \in periodset_s.snd(ev) = fst(sla) = c \,\wedge\, id\_top(ev,sla)} c\_rep\_sla\_p(c, sla, t, p)}{\#periodset_s}
\end{array}
$$

Giving an example based on the Keep Alive/PING procedure, assume the KeepAliveTimer is set to 60, then calculating $c\_rep\_sla(c, sla, t)$ will give us the reputation of the client device during the whole Keep Alive period of 60 s. In another example, based on the retransmission procedure, we assume that Nretry is set to 10. Aggregating over the $c\_rep\_sla(c, sla, t)$ values yields reputation in relation to the client's retransmissions within a 10 time-unit limit.

Finally, based on the definition of *c_rep_sla*, we can further generalise the reputation value over all relevant SLAs for a specific client, *c*, and in a particular point in time, *t*, as follows:

$$
\begin{array}{l}
=[Client, SLA]=\\
\quad c\_rep : Client \times SLA \rightarrow [0, 1]\\
\\
\quad \forall\, slaset_s : \wp(SLA) \bullet c\_rep(c, t) = \dfrac{\displaystyle\sum_{c_{rep}(c, t) = sla \in slaset_s.fst(sla) = c} c\_rep\_sla(s, sla, t)}{\#slaset_s}
\end{array}
$$

This definition gives a more general indication of how well the client device generally behaves in relation to the SLAs it holds with the server (possibly on behalf of the subscribers dealing with the server). These could include scenarios where the clients might use the Keep Alive/PING procedure to observe the liveliness of the gateway to check whether they are connected to a broker. Moreover, in the case of messages that expect a response, if the reply is not received within a certain time period, the client will be expected to retransmit this message.

The reputation model of a client in different procedures might cause different failures. Thus, as we demonstrated above the first reputation model, $c\_rep\_sla\_p$, will lead to new models with slight variations capturing this variety of failures. For example, for the case of a client's retransmission procedure, all messages that are "unicast" to the gateway and for which a gateway's response is expected are supervised by a retry timer *Tretry* and a retry counter *Nretry*. The retry timer *Tretry* is started by the client when the message is sent and stopped when the expected gateway's reply is received. If the client does not receive the expected gateway's reply during the *Tretry* period, it will retransmit the message. In addition, the client should terminate this procedure after *Nretry* number of retransmissions and should assume that it is disconnected from the gateway. The client should then try to connect to another gateway only if it fails to re-connect again to the previous gateway.

One such client reputation, is defined based on a specific *TRetry* timer:

$$
\begin{array}{l}
\boxed{\begin{array}{l}
=\!\!=\![Client, SLA, TimeStamp, TRetry]=\!\!= \\[4pt]
\quad c\_reptr\_sla\_tr : Client \times SLA \times TimeStamp \times TRetry \rightarrow [0,1] \\[6pt]
\quad \forall\, tretryset_s : \wp(Event) \bullet c\_reptr\_sla\_tr(c, sla, t, tr) = \\[4pt]
\qquad \dfrac{\displaystyle\sum_{ev\,\in\,kapingset_s.snd(ev)\,=\,fst(sla)\,=\,c\,\wedge\,id\_top(ev,sla)} \varphi(t,te)\ utility(ev,sla)}{\#tretryset_s}
\end{array}}
\end{array}
$$

For example, consider the following $sla = ((client, temperature, QoS), 1)$, then given the event $ev_{tretry} = (12:09:52, client, Publish, False, True, 1234, 2)$, it could reflect an event in the retransmission procedure. If the client does not receive a Puback message with QoS level 1 within a time period defined by the *TRetry* value, the client may resend the Publish message with the DUP flag set. When the server receives a duplicate message from the client, it re-publishes the message to the subscribers, and sends another Puback message.

Similarly, another variation of the client's reputation function may be based on the NRetry counter instead:

$$
\begin{array}{l}
\boxed{\begin{array}{l}
=\!\!=\![Client, SLA, TimeStamp, NRetry]=\!\!= \\[4pt]
\quad c\_repnr\_sla\_nr : Client \times SLA \times TimeStamp \times NRetry \rightarrow [0,1] \\[6pt]
\quad \forall\, nretryset_s : \wp(Event) \bullet c\_reptr\_sla\_tr(c, sla, t, nr) = \\[4pt]
\qquad \dfrac{\displaystyle\sum_{ev\,\in\,kapingset_s.snd(ev)\,=\,fst(sla)\,=\,c\,\wedge\,id\_top(ev,sla)} \varphi(t,te)\ utility(ev,sla)}{\#nretryset_s}
\end{array}}
\end{array}
$$

Again, for the above definition, for $sla = ((client, temperature, QoS), 2)$, and given the event $ev_{nretry} = (12:09:52, client, Publish, False, False, 1234, 1)$, it could indicate that the client should not retransmit again in the retransmission period due to the fact that QoS is set to 2 (meaning the message is guaranteed to be delivered exactly-once to the subscribers). In this case, the DUP flag must be set to False, in order to prevent a retransmission.

10.5 A reputation system architecture for MQTT

Our architecture for a reputation system for an MQTT network is composed of a *reputation monitor* and a *reputation engine*, as shown in Figure 10.1.

The architecture defines the capabilities of the various components in an MQTT network. The reputation monitor (also sometimes referred to as the proxy) will *monitor* the MQTT interactions that take place among the MQTT network components, namely the client devices, server and subscribers. Monitoring implies that the reputation monitor will issue events to the reputation engine whenever these are required after each time it has captured an MQTT communication relevant to the utility functions predefined by the consumers (possibly the subscribers). These events could represent aggregations/abstractions of data collected from such communications, in order to minimise the additional network traffic created by this process.

Once an event has arrived at the reputation engine, it is either stored for applying further aggregations/abstractions or it is used immediately to compute new updates for the various reputation values for the clients and the server. The calculations are based on the reputation models defined in the previous sections, and the updates to these reputation values are then stored in a local reputation database. In our architecture, we only consider the monitoring problem, however, it is easy to extend this architecture in the future to include a control step, where the reputation values for different participants are then used to impact/feed back into the MQTT network communications.

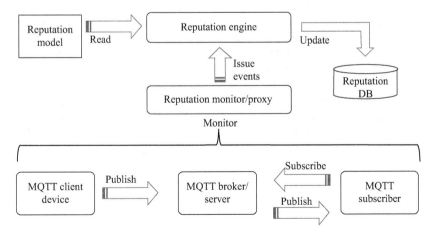

Figure 10.1   The reputation system architecture

10.6　Simulation of the model

We implemented the architecture, described in the previous section, by running a number of off-the-shelf open source tools. First, we used the Mosquitto tool [20] as the broker (server). Mosquitto is an open source MQTT broker written in the C language that implements the MQTT protocol version 3.1. The Mosquitto project is highly portable and runs on a number of systems, which made it an effective choice for our experiments.

In order to simulate the client, we used the source code for the Eclipse Paho MQTT Python client library [21], which comes with Java MQTT client API and implements versions 3.1 and 3.1.1 of the MQTT protocol. This code provides a client class, which enables applications to connect to an MQTT broker to publish messages, receive published messages and to subscribe to topics. It also provides some helper functions to make publishing one off messages to an MQTT server very straightforward. Using this library we created a set of programs that would publish and subscribe to the Mosquitto broker. Finally, to implement the monitoring function we needed to capture all the traffic between the client and the server. For this we extended the Paho MQTT *test proxy* [21], which acts as a "reverse proxy", impersonating an MQTT server and sending all the traffic on to a real broker after capturing the messages. The proxy represents a mediator between clients and the broker. By extending this proxy we were able to trace all the packets being sent and received and send monitoring information to our reputation engine in order to calculate the reputation of the client and broker.

10.6.1　Results

To begin with, we assume that the network will misbehave with regards to the messages that are exchanged among the various entities in the system. This misbehaviour is modelled as the network dropping some messages according to a predefined rate (e.g. 0%–100%). There could be other sources of network misbehaviour, such as the insertion of new messages and the repetition or modification of transmitted messages, however, for simplicity, we consider only the suppression of messages as our example of how the network could misbehave and how such misbehaviour would affect the reputation of MQTT clients and servers.

In our case, we chose the rate of successful message delivery to be in the range of 50%–100%, where 50% means that one message in every two is dropped by the network, and 100% means that every message is delivered successfully to its destination. This latter case is equivalent to the normal behaviour discussed above. There are a number of tools that can drop network packets selectively. However, we created a new tool based on the above-mentioned proxy that specifically targets disrupting MQTT flows by dropping MQTT packets. The tool allowed us to target a percentage of dropped packets and therefore calculate the reputation under a given percentage of packet loss.

Since our aim is to demonstrate, in general terms, how reputation-based trust can be obtained in an IoT system such as an MQTT network, and for simplicity, we opted to consider only one source of misbehaviour, namely message suppression, without

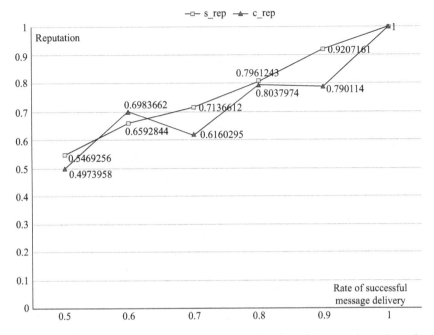

Figure 10.2 Reputation values for the Clients (c_rep) and Servers (s_rep) vs. the rate of successful message delivery

considering the other sources. Despite the fact that such sources are also interesting, they do not affect the generality of our approach.

10.6.2 Reputation results

To compute the reputation value of clients and servers, we collected events related to the QoS level agreed between the client and the server throughout the 5-min measurement window, and used the server and client reputation model proposed in Section 10.4.2 to calculate their reputation values. The QoS level monitoring is important as it is directly related to the issue of message suppression when messages are communicated over the unreliable network. In the presence of such abnormal behaviour, the reputation values of the clients and the server are shown in Figure 10.2 versus the rate of successful message delivery (0.5–1).

From this figure, we note that despite starting at low reputation levels in line with the low delivery rate of messages, these reputation values will increase reaching the optimal value of 1 when the rate of delivery of messages is 1. This optimal case represents the case of normal behaviour when every message is delivered successfully to its destination.

10.7 Related work

Reputation is a general concept widely used in all aspects of knowledge ranging from humanities, arts and social sciences to digital sciences. In computing systems, reputation is considered as a *measure* of how trustworthy a system is. There are two approaches to trust in computer networks: the first involves a "black and white" approach based on security certificates, policies, etc. For example, SPINS [22], develops a trusted network. The second approach is probabilistic in nature, where trust is based on reputation, which is defined as a probability that an agent is trustworthy. In fact, reputation is often seen as one measure by which trust or distrust can be built based on good or bad past experiences and observations (direct trust) [5] or based on collected referral information (indirect trust) [6].

In recent years, the concept of reputation has shown itself to be useful in many areas of research in computer science, particularly in the context of distributed and collaborative systems, where interesting issues of trust and security manifest themselves. Therefore, one encounters several definitions, models and systems of reputation in distributed computing research (e.g. References 5, 23, 24).

There is considerable work into reputation and trust for wireless sensor networks, much of which is directly relevant to IoT trust and reputation. The Hermes [25] and E-Hermes [26] systems utilise Bayesian statistical methods to calculate reputation based on how effectively nodes in a mesh network propagate messages including the reputation messages. Similarly TRM-IoT [27] evaluates reputation based on the packet-forwarding trustworthiness of nodes, in this case using fuzzy logic to provide the evaluation framework. Another similar work is CORE [28] which again looks at the packet forwarding reputation of nodes.

Our approach differs from the existing research in two regards: first, the existing reputation models for IoT utilise the ability of nodes to operate in consort as the basis of reputation. While this is important in wireless sensor networks, there are many IoT applications that do not utilise mesh network topologies and therefore there is a need for a reputation model that supports client–server IoT protocols such as MQTT. Second, the work we have done evaluates the reputation of a reliable messaging system based on the number of retries needed to successfully transmit a message. Although many reputation models have been based on rates of packet forwarding, the analysis of a reliable messaging system (like MQTT with QoS > 1) is different as messages are always delivered except in catastrophic circumstances. Therefore we looked at the effort and retries required to ensure reliable delivery instead. We have not seen any similar approach to this and consider this the major contribution of the chapter.

10.8　Discussion

To conclude, we defined in this chapter a model of reputation for IoT systems, in particular, for MQTT networks, which is based on the notion of utility functions. The model can express the reputation of client and server entities in an MQTT system at various levels, and in relation to a specific issue of interest, in our case the QoS level of the delivery of messages in the presence of a lossy network. We demonstrated that it is possible, using off-the-shelf open source MQTT tools, to implement an architecture of the reputation system that monitors the MQTT components, and we showed that the experimental results obtained from running such a system validate the theoretical model.

Future work will focus on adapting the reputation model and its architecture and implementation to other IoT standards, e.g. the Advanced Message Queuing Protocol (AMQP) [29], the Extensible Messaging and Presence Protocol (XMPP) [30], the Constrained Application Protocol (CoAP) [31] and the Simple/Streaming Text Oriented Messaging Protocol (STOMP) [32]. We also plan to consider other issues of interest when calculating reputation where satisfaction is not necessarily a binary decision, e.g., the quality of data generated by client devices and the quality of any filtering, aggregation or analysis functions the server may apply to such data in order to generate new information to be delivered to the consumers. Further, we intend to apply Bayesian statistics to the results to improve the probabilistic calculation of the reputation values.

Some other interesting, though more advanced areas of research, include the strengthening of the model to be able to cope with malicious forms of client and server behaviour, e.g., collusion across such entities in order to produce fake reputation values for a targeted victim, and a study on the welfare of IoT ecosystems based on the different rates of the presence of ill-behaved and well-behaved entities in the ecosystem, and how variations in the presence ratio of such entities would lead to a notion of reputation reflecting the wider ecosystem.

References

[1]   D. Locke, *MQ Telemetry Transport (MQTT) V3.1 Protocol Specification* IBM and Eurotech, (2010).

[2]   A. E. Arenas, B. Aziz, G. C. Silaghi, "Reputation management in collabora-tive computing systems", *Security and Communication Networks* 3 (6) (2010) 546–564.

[3]   G. C. Silaghi, A. Arenas, L. Silva, "Reputation-based trust management systems and their applicability to grids", Tech. Rep. TR-0064, Institutes on Knowledge and Data Management & System Architecture, CoreGRID – Net-work of Excellence (February 2007). Available from: http://www.coregrid.net/mambo/images/stories/TechnicalReports/tr-0064.pdf, accessed: 2016-04-04.

[4]   Merriam-Webster. Available from: http://www.merriam-webster.com, accessed: 2016-03-11.

[5]   A. Jøsang, R. Ismail, C. Boyd, "A survey of trust and reputation systems for online service provision", *Decision Support Systems* 43 (2) (2007) 618–644.

[6]   A. Abdul-Rahman, S. Hailes, "Supporting trust in virtual communities", in: *HICSS'00: Proceedings of the 33rd Hawaii International Conference on System Sciences*, Vol. 6, IEEE Computer Society, Washington, DC, USA, 2000.

[7]   T. Grandison, M. Sloman, "A survey of trust in Internet applications", *IEEE Communications Surveys and Tutorials* 3 (4) (2000) 2–16. Available from: http://pubs.doc. ic.ac.uk/TrustSurvey/, accessed: 2016-04-04.

[8]   D. Gambetta, "Trust: making and breaking cooperative relations", Department of Sociology, University of Oxford, 1988, Ch. Can We Trust Trust?, pp. 213–237, Available from: http://www.sociology.ox.ac.uk/papers/gambetta 213–237.pdf, accessed: 2016-04-04.

[9]   B. Yu, M. Singh, "An evidential model of distributed reputation management", in: *AAMAS'02: Proceedings of the First International Joint Conference on Autonomous Agents and Multiagent Systems*, ACM Press, New York, NY, 2002, pp. 294–301. doi:http://doi.acm.org/10.1145/544741.544809, accessed: 2016-04-04.

[10] D. Fudenberg, J. Tirole, *Game Theory*, MIT Press, Cambridge, MA, USA, 1991.

[11] C. Dellarocas, "How often should reputation mechanisms update a trader's reputation profile?", *Information System Research* 17 (3) (2006) 271–285.

[12] P. Resnick, R. Zeckhauser, *Advances in Applied Microeconomics*, Vol. 11, Elsevier, Amsterdam, 2002, Trust Among Strangers in Internet Transactions: Empirical Analysis of eBay's Reputation System.

[13] G. Zacharia, P. Maes, "Trust management through reputation mechanisms", *Applied Artificial Intelligence* 14 (9) (2000) 881–907.

[14] M. Singh, B. Yu, M. Venkatraman, "Community-based service location", *Communications of the ACM* 44 (4) (2001) 49–54. doi:http://doi.acm. org/10.1145/367211.367255.

[15] A. Banks, R. Gupta, *MQTT Version 3.1.1* OASIS Open (2015).

[16] K. Birman, T. Joseph, "Exploiting virtual synchrony in distributed systems", *SIGOPS Operating Systems Review* 21 (5) (1987) 123–138.

[17] B. Aziz, G. Hamilton, "Reputation-controlled business process work flows", in: *Proceedings of the Eighth International Conference on Availability, Reliability and Security*, IEEE CPS, 2013, pp. 42–51.

[18] B. Aziz, G. Hamilton, "Enforcing reputation constraints on business process work flows", *Journal of Wireless Mobile Networks, Ubiquitous Computing, and Dependable Applications (JoWUA)* 5 (1) (2014) 101–121.

[19] T. D. Huynh, N. R. Jennings, N. R. Shadbolt, "An integrated trust and reputation model for open multi-agent systems", *Autonomous Agents and Multi-Agent Systems* 13 (2) (2006) 119–154.

[20] "Mosquitto: An open source MQTT v3.1/v3.1.1 broker". Available from: http://mosquitto.org/, accessed: 2016-03-11.

[21] Paho. Available from: http://www.eclipse.org/paho/, accessed: 2016-03-11.

[22] A. Perrig, R. Szewczyk, J. Tygar, V. Wen, D. E. Culler, "Spins: security protocols for sensor networks", *Wireless Networks* 8 (5) (2002) 521–534.

[23] K. Fullam, K. Barber, "Learning trust strategies in reputation exchange networks", in: *AAMAS'06: Proceedings of the Fifth International Joint Conference on Autonomous Agents and Multiagent Systems*, ACM Press, Hakodate, Japan, 2006, pp. 1241–1248.

[24] G. C. Silaghi, A. Arenas, L. M. Silva, "Reputation-based trust management systems and their applicability to grids", Tech. Rep. TR-0064, Institutes on

Knowledge and Data Management and System Architecture, CoreGRID – Network of Excellence (February 2007).

[25] C. Zouridaki, B. L. Mark, M. Hejmo, R. K. Thomas, "Hermes: A quantitative trust establishment framework for reliable data packet delivery in MANETs", Journal of Computer Security 15 (1) (2007) 3–38.

[26] C. Zouridaki, B. L. Mark, M. Hejmo, R. K. Thomas, "E-hermes: A robust cooperative trust establishment scheme for mobile ad hoc networks", *Ad Hoc Networks* 7 (6) (2009) 1156–1168.

[27] D. Chen, G. Chang, D. Sun, J. Li, J. Jia, X. Wang, "TRM-IoT: a trust management model based on fuzzy reputation for internet of things", *Computer Science and Information Systems* 8 (4) (2011) 1207–1228.

[28] P. Michiardi, R. Molva, "Core: a collaborative reputation mechanism to enforce node cooperation in mobile ad hoc networks", in: *Advanced Communications and Multimedia Security*, Springer, Portorož, Slovenia, 2002, pp. 107–121.

[29] S. Vinoski, "Advanced message queuing protocol", *IEEE Internet Computing* 10 (6) (2006) 87–89.

[30] XMPP Standards Foundation. Available from: http://xmpp.org, accessed: 2016-03-11.

[31] Z. Shelby, K. Hartke, C. Bormann, "Constrained Application Protocol (CoAP)", draft-ietf-core-coap-18. Available from: https://tools.ietf.org/html/draft-ietf-core-coap-18, accessed: 2016-03-11.

[32] "STOMP: the simple text oriented messaging protocol". Available from: https://stomp.github.io, accessed: 2016-03-11.

Index

Printed in the USA
CPSIA information can be obtained
at www.ICGtesting.com
JSHW011509221024
72173JS00005B/1252